PRAISE FOR *CITY ON FIRE* BY BILL MINUTAGLIO

"Incendiary reading. . . . A harrowing mosaic about a blaze during a time of racial divisions and environmental plundering . . . evocatively told. [Minutaglio's] hard-edged prose brands scores of images on readers' minds. . . . The book vividly details the carnage as well as some acts of heroism and selflessness." —*Publishers Weekly*

"A poignant present-tense account of the hours before, during, and after the explosion, bringing to life the horror, pain, and bravery of the people of Texas City. . . . This terrible story deserves this passionate retelling." —*Library Journal* (starred review)

"A complex portrait of an industrial city and the powers that created and controlled it. . . . *City on Fire* will stand on its own as one of the finest books ever written about Texas." —*Texas Observer*

"[Bill Minutaglio] interweaves heart-wrenching personal experiences with the collective story of a town attempting to recover from a monumental tragedy. . . . Reminiscent of New York City's rise from the ashes after September 11, the chronicle of Texas City's devastation and resurrection will strike a chord with contemporary readers." —*Booklist*

"*City on Fire* . . . is a remarkable re-creation. It provides the reader with perspective from every corner of a town balkanized by race, ethnicity, religion, and class. Minutaglio's technique is both novelistic in its attention to character development and cinematic in its rapid shifts from scene to scene. . . . A terrific nonfiction work that has the narrative force of an adventure novel." —*Washington Post*

"Shrewdly focused on a group of key figures, Minutaglio's account provides a highly personalized portrait of the tragedy that struck Texas City. . . . Minutaglio conveys a punchy, noirish sense of the period." —*Kirkus Reviews*

"Riveting . . . historical reporting at its best. Minutaglio writes with such skill, one can almost feel the gritty, humid heat of Texas and smell the emissions of the nearby refineries. . . . A story about courage, bravery, and a painful quest for justice." —*Tucson Citizen*

"[Minutaglio]'s meticulous research and brilliantly graphic narrative style make *City on Fire* a memorable and lasting book."

—*Dallas Morning News*

"All who read *City on Fire* will find it impossible to forget what happened in Texas City in 1947. Bill Minutaglio's re-creation of the cataclysm, and its tense and scandalous sequel, is storytelling at its best. . . . An important piece of American history."

—Harold Evans, author of *The American Century*

"My father was a natural gas engineer and was at Texas City on the April morning when it blew up in 1947. *City on Fire* is a totally accurate narrative about the events of that tragic day. The story of Texas City is the story of the Cold War, of greed, lust, environmental degradation, racism, and the unbridled power of corporations. But this book is not simply a history of past events. It's a warning about the world in which we live, about politicians who serve the interests of the corporations rather than their constituencies. But it is also a story about decency and compassion and spiritual resolve and courage."   —James Lee Burke, author of *Jolie Blon's Bounce*

"The hidden or forgotten tragedy is often the worst—leaving the victims doubly bereft by being forgotten. *City on Fire* sheds a poignant light on an important event in modern American history. Minutaglio has produced an exquisitely readable and painstakingly reported work of narrative nonfiction."

—Gail Sheehy, author of *Hillary's Choice* and *The Silent Passage*

"The Texas City disaster was one of the biggest news stories of its time. This stunning piece of historical re-creation restores this tragic and landmark event to its rightful place in the national memory."

—Dan Rather, *CBS News*

"*City on Fire* is history at its best, at once thrilling and illuminating. The story of ambition, hubris, tragedy, and bravery . . . is as timeless today in all of America as it was back in Texas more than half a century ago."

—David Maraniss, author of *When Pride Still Mattered*

John Anderson

About the Author

BILL MINUTAGLIO is the author of four books, including the highly acclaimed biography of President George W. Bush, *First Son: George W. Bush and the Bush Family Dynasty*. He has won numerous national awards and has written for the *New York Times, Talk, Outside,* the *Los Angeles Times,* and the *Bulletin of Atomic Scientists*. Bill Minutaglio lives with his wife, the choreographer Holly Williams, and their two children in Austin.

City on Fire

BILL MINUTAGLIO

The Explosion
That Devastated a Texas Town and Ignited
a Historic Legal Battle

Perennial

An Imprint of HarperCollinsPublishers

A hardcover edition of this book was published in 2003 by HarperCollins Publishers.

CITY ON FIRE. Copyright © 2003 by Bill Minutaglio. All rights reserved. Printed in the
United States of America. No part of this book may be used or reproduced
in any manner whatsoever without written permission except in the case of
brief quotations embodied in critical articles and reviews. For information
address HarperCollins Publishers Inc., 10 East 53rd Street, New York, NY 10022.

HarperCollins books may be purchased for educational, business, or sales promotional
use. For information please write: Special Markets Department, HarperCollins
Publishers Inc., 10 East 53rd Street, New York, NY 10022.

First Perennial edition published 2004.

Designed by Sarah Maya Gubkin

The Library of Congress has catalogued the hardcover edition as follows:

Minutaglio, Bill.
    City on fire: the forgotten disaster that devastated a town and ignited a landmark
legal battle / Bill Minutaglio.—1st ed.
        p.   cm.
    ISBN 0-06-018541-4
        1. Fires—Texas—Texas City—History—20th century.   2. Disasters—Texas—
Texas City—History—20th century.   3. Industrial accidents—Texas—Texas City—
History—20th century.   4. Texas City (Tex.)—History—20th century.   5. High
Flyer (Ship).   6. Grandcamp (Ship).   7. Wilson B. Keene (Ship).   8. Texas City
(Tex.)—Biography.   I. Title.

F394.T4 M56   2003
976.4'139—dc21

                                                                        2002069064

ISBN 0-06-095991-6 (pbk.)

        04   05   06   07   08   ❖/RRD   10   9   8   7   6   5   4   3   2   1

To Holly

A visionary light settled in her eyes. She saw the streak as a vast swinging bridge extending upward from the earth through a field of living fire. Upon it a vast horde of souls were rumbling toward heaven.

—Flannery O'Connor, from "Revelation," in her anthology
Everything That Rises Must Converge

Introduction

Before international terrorists used ammonium nitrate in the first attack on the World Trade Center in 1993, before Osama bin Laden's network ordered the 1998 ammonium nitrate bombing of U.S. embassies, before Timothy McVeigh's 1995 ammonium nitrate bombing of a federal building in Oklahoma . . . there was a far deadlier explosion in a small American city. It was also an ammonium nitrate explosion—but it was three hundred times more powerful than the one unleashed by McVeigh.

It claimed more lives on American soil than any other man-made disaster in the twentieth century. It was the greatest industrial tragedy in the history of the most industrialized country in the world.

On a beautiful spring day in 1947, the hardworking and perfectly American town of Texas City, Texas, almost disappeared from existence. Everyone in the small city was forced to face an endless series of converging, disparate forces.

Thousands of people were killed or wounded, a fire department was erased, planes fell from the sky, oceangoing freighters vanished. The most powerful people of the era—the president, Supreme Court justices, military commanders, Hollywood superstars, FBI directors—were drawn into the plot.

And then the patriotic residents of the small city found themselves becoming among the first to contemplate something equally impossible—the belief that the blood of American citizens could be found on the hands of their leaders.

The Texas City Disaster, as it came to be known, would have lingering effects for millions of us.

It would set legal standards for determining if our elected officials have been horribly negligent in their duties to protect and serve the American people. It would redefine the entire way federal, state, and local officials respond to the most massive emergencies—including the 9/11 tragedies.

But, in time, many other Americans would cease to remember how the ordinary heroes in Texas City found themselves hurtling toward modern science's deadly miracles—and toward the international schemes of the world's leaders.

No one will ever know how many men, women, and children died in Texas City. Some say six hundred, some say seven hundred or eight hundred—so many simply vanished that no one knows the real count. There were five thousand injured. Over two thousand homeless. Businesses, defense plants, refineries, houses, churches, planes, cars, trains, and huge ships were destroyed. The initial property damage was close to $4.5 billion. The lingering costs went much higher.

At the heart of the disaster was that once-seemingly magical compound called ammonium nitrate.

At the dawn of the twentieth century, the world's brilliant scientists came frantically to the conclusion that the Earth's natural nitrogen cycle would no longer be able to support life on the planet.

Influential chemists raced to their laboratories. They uncorked fabulous experiments with nitrogen, until God's own coda seemed to be playing and startling, life-giving nitrogen compounds could be created in massive proportions.

The Nobel Prize was dutifully awarded to one one of those chemists, in grateful recognition that his work had saved mankind—that his compounds could be used to feed millions of starving people.

But it also became quickly clear that the thrilling chemistry had a profound dark side.

Those exact life-saving compounds could also be used to make the deadliest explosives—ones that could conduct the great wars, destroy office towers, and level cities.

It was a horrific paradox—the chemical miracles could be enduring nightmares.

In time, the world would become addicted to both possibilities:

The United States would use millions of pounds of the nitrogen compounds to bomb its enemies into submission—and it would use billions of pounds of the same compounds to grow crops in vulnerable, strategically important countries around the world.

Some studies show that American farmers now use about 17 billion pounds each year of the nitrate fertilizers. A federal study not long ago concluded that there is no pragmatic way to make ammonium nitrate safer "without seriously affecting its use as a fertilizer."

It is a cruel irony that the good people of Texas City know better than anyone else in the country.

That irony first began to become clear for me one morning on a day that promised to emerge as beautiful as that one in 1947.

Exactly twenty years ago, I stood in the inky dark on the Texas City waterfront, fishing and waiting for the sun. When dawn finally arrived, it knocked my breath away. As the sun rose, there was the massive silhouette of a rusted ocean-bound freighter—and behind it, endless miles of smoking, towering, dull-gray petrochemical plants.

It was like seeing something ominous rising up out of Texas City's past.

From that day on, I began years of research into what really happened when Texas City became a city on fire.

More than two hundred people were interviewed for this book. More than thirty thousand pages of trial transcripts, federal inquiries, congressional committee hearings, and reports housed in the United States National Archives were examined. Original police logs, depositions, and deathbed statements were examined. Eyewitnesses who were within yards of the explosion were interviewed for the first time. International maritime experts were consulted. Thousands of pages of out-of-print newspapers, magazines, and medical journals were unearthed. Thousands more pages of FBI documents were ordered under the Freedom of Information Act. The archives of the Catholic Church were consulted.

The same rules used in my biography of President George W. Bush were applied. Exact quotes from transcripts, interviews, previously unpublished material, and depositions are in quotation marks and sometimes italicized for emphasis. Other internal and external dialogue, built from those same sources but not gleaned from firsthand accounts, is italicized but not in quotes. As we did with the biography of President Bush, I worked with my researchers to keep this work as bias-free as possible.

Father Roach, the blessed soul who cared so much for everyone from every corner of Texas City, was brought to life by the vivid letters, extensive biographies, correspondence, photographs, and archival records generously provided by Lisa May, the brilliant archivist with the Catholic Diocese in Galveston, Texas—and by the intimate, detailed memories of several warm and good-hearted priests, nuns, friends, coworkers, and parishioners who faithfully cataloged or recalled the statements, actions, and conversations of this saintly man and his family. Among the dozens of kind people across America who had specific details of Bill Roach's life were these good folks: Rev. James Vanderholt, Sister William, Rev. Frank Doremus, Monsignor Jamail, Rev. Thomas Culhane, Thelma Avant, Elva Rogers, Doris Pire, and Bernice Smith; their sharing of their personal memories and connections to Father Roach were supplemented by personal letters written to Father Roach by the Bishop of Galveston, as well as articles, obituaries, essays, and reminiscences about him in *Southern Wind* and other religious publications.

And now, after my years of obsession, one thing still haunts me about what happened that awful day in Texas City. One thing still rises up like that rusted ship did one dawning morning so many years ago:

The stories of ordinary, heroic Americans racing to an apocalypse should never be forgotten.

Bill Minutaglio
Austin, Texas
Winter 2002

City on Fire

TEXAS CITY POLICE DEPARTMENT

Radio Dispatcher's Log
Wednesday, 4/16/47

05:31—Time test

06:00—Time test

06:45—Car #752 out of service and at police station

07:49—Radio frequency checked by Chief of Police W. L. Ladish

08:15—Car #752 out of service and at Clark's Department Store

08:35—Working the traffic at Dock Road and Third Street

08:37—Report of fire: Terminal Docks

08:40—Car #753 reports cotton burning on freighter

09:19—EMERGENCY

11:15—MORE STRETCHERS FOR DEAD

The Priest

APRIL 7, 1947

Texas City

His HEART has been racing lately, and his thin hands have been shakily reaching for the first of his two daily packs of Old Gold smokes. Each night over the last two weeks it's been the same thing. It's something numbing and far beyond weariness, beyond depression. It's like stepping through a door he's opened a million times before, and falling, drowning, into a tunnel of black cotton.

Hours ago, after he had finished dinner at the home of some friends, he had led his twin brother into the moist evening. The fronds on two raggedy palms down by Bay Street were flapping like the arms of a straw man. A mile out in moonlit Galveston Bay, the swaying spotlights from the shrimp boats reflected on the water as if they were fireflies moving over a mirror. Bill Roach leaned against his crinkled, rusted Ford and stared into the open face of his twin brother, Johnny. They had done a lot of traveling together. They had been inseparable for years, coming to Texas, going back to their hometown of Philadelphia—locked in endless road trips across America.

Bill finally tells Johnny that he is going to die: *"I'm not quite resigned to die yet. I still have a lot I'd like to do. But that's what God wants and I have to accept it . . . surrender."*

It was, of course, like looking into his own damned face. They were almost perfectly identical. Bill let his words settle in. Johnny, unblinking behind his glasses, didn't make a sound or move a muscle. Bill could see his brother holding tight. Finally, Johnny exhaled and said he'd try to understand, even if he couldn't.

Bill said there was more: *"It's not just me. A lot of people are going to die. There's going to be blood in the streets."*

A TUGBOAT CAPTAIN plowing through the pewter-colored waters off of Texas City says the waterfront "looks like an ominous Oz"—a gunmetal fortress of towering, riotously intertwined pipes, catalytic cracking furnaces, steam superheaters, domed oil storage tanks, catwalks, condensers, and two-hundred-foot-high funnels shooting twenty-foot-long gas flares into the sky. All of it is linked to long lines of tin-roof warehouses, rusted railroad cars, and the concrete-lined docks—and all of it is coursing with the liquid lifeblood of Texas City: millions of barrels of oil, gas, benzol, propane, benzene, kerosene, chlorine, styrene, hydrochloric acid, and a necromancer's trough of every other petrochemical imaginable.

For decades, this billion-dollar stretch of oil refineries, oil tank farms, and chemical plants has been turned into one of the most lucrative, strategic petrochemical centers in the world by the Rockefellers, Howard Hughes, and even the far-flung members of the Bush family.

Forty miles southeast of Houston, ten miles north of Galveston, Texas City has been carved out of an isolated, unforgiving stretch of the Gulf of Mexico coastline. The yawning, splotchy sky often looks like stained concrete. The pregnant humidity rubs against the skin like a heated stranger. Clouds of mosquitoes puff up out of mushy ditches.

And twenty months earlier, Bill Roach couldn't have been happier to be anywhere else on earth.

He had called Johnny to tell him that he had just sliced open a letter from the Catholic bishop in Galveston. It contained the orders assigning Bill to Texas City. Roach was ecstatic. It was as if his entire life had been a process of preparation for coming to this hidden piece of the United States.

Texas City was a vital, heavily guarded defense-industry town during every phase of World War II. Half the almost twenty thousand people who lived there were making gas for military vehicles, aviation fuel for American planes, tin for a thousand different military uses, tons of synthetic rubber to keep tanks, Jeeps, and troop trucks rolling. Texas City was a frontier city run by the military, oil companies, metal companies, and chemical companies who were doing billions of dollars of work for the war effort.

When the war ended, the town never slowed down—it simply became another staging area for the next war, the one that would be called the Cold War.

Years earlier, the Chamber of Commerce had come up with a slogan: *"Texas City: Port of Opportunity."* Lately, the chamber had been using another slogan: *"Texas City: Heart of the Greatest Industrial Development in the Country."* The local newspaper, the *Sun*, was using variations of both in its masthead.

In the last seven years Texas City has tripled in size, and every day more strangers are setting foot there from every corner of the country. When those newcomers first turn off the Houston highway and steer east to Texas City and its waterfront, they can see through their bug-splattered windshields that the place has been hammered into existence without regard to aesthetics.

They are arriving at an American town with few pretensions. Texas City lacks the theaters, operas, greenbelts, promenades, burnished mansions, gazebos, seaside parks, public monuments, grand libraries, and fine restaurants that are found in Dallas, Houston, and New Orleans. It is a working-class coastal town, and most everyone knows that the massive amounts of money generated there rarely stay in the city. People are paid a living wage, often because they belong to a union, but the billion-dollar industrial zone has barely been taxed. Union Carbide, Amoco, and other international industries clustered into that metal world fanning out from the waterfront are located just outside the old city limits set up by the founding fathers—and they never generated tax revenues to put up streetlights, to open more schools, to move the library out of the back of City Hall, to suck the mosquito ditches, to pay for a fireboat.

But Texas City has steady paychecks.

At the downtown station, Roach runs into discharged GIs jumping off out-of-state buses and looking for jobs. When Roach follows the narrow road to the Houston highway, there are freshly arrived high school graduates who've just hitchhiked in from dusty, going-broke towns as far away as El Paso. They're all gravitating toward a knot of young men in dirty white T-shirts hanging out in front of the Showboat Drugstore on Sixth Street.

A newcomer from Dallas named Tommy Burke is there, blowing smoke rings with his Lucky Strikes, flirting with the Monsanto Chemical secretaries on their lunch breaks, and swapping leads on who's hiring that week: *Republic Oil had a little butane explosion, killed a couple of guys; they need a construction crew, and they're gonna probably hire some more people for their fire crew.*

Texas City is the fourth-largest port on the Gulf of Mexico. It's one of the fastest-growing cities in the United States. The Seatrain line has made Texas City a favored point of departure for New York. Humble Oil, later to become Exxon, has a giant presence. So does Amoco. There are oil tank farms owned by the billionaire Texas oil sheiks. Monsanto and Union Carbide are hiring every day. The tin smelter is the largest in the world. Tons of cotton, peanuts, and molasses need loading.

And at night, Roach watches as the mates from French freighters, Dutch tankers, and New Orleans–based cargo ships fan out into the narrow oyster-shell alleys leading away from the gumbo soil around the concrete docks. The only cab company in Texas City has drivers ready to take them up to one of the red-light joints . . . someplace where the cabbies have a connection. When they saunter into the cool darkness of Tillie's or Jeanette's, the seamen are shoulder to shoulder with the sweaty footsoldiers in the blue-collar battalions: pipe welders, trench diggers, heavy construction crews, pipe fitters, boilermakers, draftsmen, furnace operators, heavy machinists, wharf crews, turbine operators, and "sample men"—ballsy laborers whose sole, thankless job is to climb fifty feet up in the air, hover over enormous petrochemical cauldrons, and dip bottles into the roiling soup in order to collect noxious, toxic samples for the chemists.

Sometimes, too, there are the corporate emissaries sent by the Rockefellers, by Monsanto's Edgar Queeny, by Howard Hughes, and by Humble Oil director Jesse Jones—the men in expensive wool suits from Neiman Marcus, the ones who did their paperwork and whoring in Texas City but preferred finally to fall asleep twelve miles away in the fine hotels of languid, tropical Galveston.

On his rounds down by the rowdy waterfront, Roach has developed a theory about it all: *Everyone who lives here is a refugee steered to Texas City by a rising tide.*

Roach and and his friend Mayor Curtis Trahan have no doubt heard the running gag. It revolves around how in holy hell each and every one of those dreamers found himself in Texas City—as if it were the last stop on the railroad heading west, as if it were the location of the latest gold rush.

"What the hell brought you here?"

The joke answer is always the same, especially when you sucked in a deep breath of that overpowering, raw oil-and-chemical-and-seaside stench that seemed to coat the inside of your lungs:

"Shit, that's an easy answer. It's the stink. The stink of money."

And now Bill Roach feels smothered by visions of blood coursing down the streets of Texas City.

The Voyage

April 7, 1947

FROM THE deck of his bruised ship, a veteran engine stoker named Pierre Andre is staring into an impenetrable wall of fog. There is a chemical tang to the air, the taste of a penny. The humidity is thick, as if you could pinch a piece of air and bring it to your lips. When Andre and the other mates step across the sheen-coated bridge, it is like struggling through an invisible waist-high field of damp reeds. Andre and the other men have been away from their families for four months.

They are aboard the S.S. Grandcamp, a 7,176-ton freighter that is the color of rust and shark's skin, docked in the narrow leg of the Houston Ship Channel. Sailors are scrambling across the fifty-six-foot-wide deck, trying to ready the vessel in case that wavering, noxious fog can finally be breached. Since Christmas, the ship has had a grinding, cargo-hauling tour that began in New York, then south to Newport News, Virginia, for a load of coal, and across the treacherous North Atlantic seas to Cherbourg and Rouen in France. Then it was on to Antwerp, Belgium, and, immediately, a recrossing of the Atlantic to Venezuela. That port of call was followed by a sprint north to Havana and, finally, special orders to pick up one last cargo load in Texas. Word had filtered down to the sailors about their bad luck—the French shipping company that owned the Grandcamp had originally wanted another one of its vessels to make the run to Texas but had ordered the Grandcamp to dock there at the last instant. By April, the Grandcamp was plowing northwest, back into American waters, destined first for Houston and then finally a brutishly unattractive port on the Gulf Coast called Texas City. There it will take on its final cargo—several thousand tons of precious ammonium nitrate.

The forty sailors are old salts. Some of them have been going to sea since they were teenagers. Like all of them, Andre lives modestly when he's waiting for the next phone call or message ordering him back to a ship that needs some men, back to foreign ports like exotic-sounding Texas City. Andre and his family have a small fisherman's bungalow in Cotes d'Armor. It is like other hardscrabble seaside places around the world, a place of unpaved streets where everyone is accustomed to days that stretch into months, where the only echoing voices are from the women and children . . . where the men won't be back home until the seasons change twice.

When he was getting ready to leave his family and home this time, Andre had sat in a chair to finish putting on his shoes. His fourteen-year-old daughter stared at him, searing his image into her memory. She saw tears begin to slide down her father's cheeks.

Outside there was the same resigned ritual: the women and the children watching their sailors trudging down the muddy streets, walking to the taxis that would deliver them to the big ships.

Andre's wife and daughter watched him walk away from the house. And when they looked down at the pockmarked soil, they could see a long set of perfect imprints from Andre's boots, disappearing into the distance.

For days, the tracks remained intact. It seemed as if Andre's tracks would be there forever. They stayed there so long that his wife and daughter began to believe there was a profound, prophetic reason those tracks never eroded.

They assumed it was some sort of disturbing sign, and they made sure never to erase them.

The Priest

APRIL 8, 1947

JOHNNY, FRIENDS WOULD SAY, NEVER brings it up again with his brother.

The night Bill told him he was going to die, the brothers hugged and said good-bye. Johnny, usually fearless, headed up the oil-coated, narrow highway to the rising skyline of Houston and felt he had probably seen his brother for the last time.

The next day, Bill is out on Sixth Street, making his rounds. He starts at Lucus's Café, then heads over to City Hall to badger his friend Mayor Curtis Trahan about when the hell they are going to push forward their plan to get running water—to get anything—for all the blacks and Mexicans jammed along the waterfront.

But people can see a change. Bill Roach's skin is no longer ruddy; it is ashen. He has a dusty pallor that's beyond the usual one that coats a chain-smoker. An oil refinery worker spots him outside the longshoreman's union hall and says he is, as always, energized. But this time Roach looks nervous; he's moving at a discordant tempo. He's jangled, edgy, unsettled—and it's all the more evident on a normally smooth operator like Roach.

The sailors, longshoremen, and dock jockeys lingering in the cheap glow coming from The Stumble Inn, The Longhorn, and all the whorehouses and dives close to the docks usually love to see him striding through the petrochemical fog with a lit cigarette dangling from his thin lips. When he whirls into a room, preceded by the glow of another Old Gold, friends and enemies snap to him like metal shavings popping toward a magnet. Past battles with diabetes have given him a slight limp and a bouncy, rolling stride that people chalk up to his urgent nature. It could be at Norris's Café, at the El Charro Restaurant, at Trahan's City Hall office, in the hallways of the company that ran the railroads and the docks. Roach is usually unavoidable, outspoken on everything that shapes the quality of life in Texas City.

He's usually got plenty of Irish.

He's normally like something out of a Bing Crosby movie, the one in which Crosby plays the battling, insouciant priest sliding through tight spots with a wink to the girls and a nod to the boys. Usually, when Bill Roach rolls down Third Street, those old salts and the other blue-collar knuckleheads spending their refinery paychecks on Grand Prize beer forget he's wearing a clerical collar. Usually, when it comes to Roach, it's all smiles, slaps on the back, and hurrahs. He has the barnyard bounce of a good bantamweight boxer. He knows jokes, and his voice, edged with a streetwise Philadelphia accent, has a habit of lingering long after he is gone.

He's right there with you, you know?

But these last two weeks, even the bleary-eyed just walking out the back door of Tillie's stale-smelling club can see it. Roach is off his rhythm. And a handful of his many admirers suspect something is very wrong. A devoted woman in his parish, Bernice Smith, says it appears as if something is literally consuming Bill Roach, eating the priest alive.

"He looks emaciated."

Two weeks ago, Smith had been driving by the church and she saw Roach sitting on the steps. It was eight P.M. She rolled up alongside him and good-naturedly yelled out the car window:

"What are you doing sitting out here?"

Roach stared back and said:

"I feel like I'm sitting on a keg of dynamite—and I have no idea what to do about it."

Smith didn't know what to say. She had no idea what the hell he was talking about. Roach, she thought, looked scared. That wasn't like him. The Roach boys were usually tougher than steel.

Yesterday, Smith and her third-grader, Rita, drove to the church. Rita had talked her mother into stopping at St. Mary's to light a candle. As the mother and daughter stepped up to the altar, they could hear the swish of long robes and the echo of footsteps on the cool tile. Roach was alongside them, ominously putting his right hand on the eight-year-old girl's forehead:

"Rita, say a prayer for me."

ROACH'S HANDS ARE still doing a Saint Vitus' dance, and his lit cigarette is like a mad, bouncing firefly. Roach has never had the shakes before. He has never had trouble sleeping. Now, he thinks it might be good to get the hell out of his damned small bungalow, to go down to the docks, to walk over the bleached-looking shells that are used to pave the alleys, to spend some time shooting the shit with the union guys and the sailors.

Every morning, before the sun bleeds through the hideous smog, Roach knows that down by the whorehouses, half the distance to the waterfront from his bungalow, the last stumbling drunken longshoremen, shipmates, and prostitutes are snaking out onto the muddy avenues and the soupy ditches, aiming for their one-room apartments and metal ship's bunks.

They'll pass through a ten-block neighborhood of unpaved streets and clapboard shacks. This neighborhood, at the hem of the factories and refineries on the waterfront, is carved into two sections: one for blacks and the other for Mexicans. The people who live on the Negro side sometimes call it The Bottom. The people who live on the Mexican side would someday call it El Barrio; others call it Spanish Town.

Roach has always considered himself a rational, empowered man—as rational as any of the white-collar chemists and scientists who have been sent to oversee operations at the Texas City plants. He can talk to anyone—saints and sinners, union guys and company presidents. But he also sees himself as a student of the raw human condition; as a social scientist, somebody who takes the temperature of each city he is assigned to work in. He has deliberately set himself up in Texas City—

knowing that it is one of America's most vital industrial zones, knowing that it is a place where the rich always get richer and the blue-collar factory workers simply scramble to stay stable.

Roach had come to Texas City to correct a few things.

His specific techniques were spelled out yesterday when Roach marched into the well-worn lobby of the *Texas City Sun* on Sixth Street and told the wary editors he was going to pay for a large, prominent space in the paper. His scathing open indictment of Texas City appeared in the Tuesday edition, word for incendiary word. It was even more intense than the speeches Roach delivered at the public hearings in the second-floor meeting room of the old City Hall.

"Texas City Absorbs Industry or Industry Absorbs Texas City?" was the giant headline above the text:

> *While Hitler was sealing the fate of the future of Germany, our American Government was directing and building a greater America into a stronger nation. The differential was, one was motivated by avarice, selfishness, and greed, while the other was based on Jefferson's Democracy, that is, giving freedom, protection, and help to the poor and depressed, which is the foundation of America's greatness. While one was attracted by the hideousness and the ugliness of mud, the latter saw the beauty of the stars.*
>
> *It is needless to point out to the citizens the primitive and sordid conditions under which we are living. For a sewage disposal plant we are using our front yard, that is, the bay. For a garbage incinerator, we dig large trenches on the prairie, leaving the community open to the potential danger of epidemics. ABOMINABLE! Yet we pride ourselves on being modern. Why, such conditions do not exist in the most primitive countries of South America!*
>
> *Our Mexican and Negro sections are a disgrace . . . both of these areas are over-populated . . . two and three families are living in one or two rooms . . . outdoor toilets, open sewage. . . .*

The town is encircled by a "steel band" held firmly by a few selfish individuals.

FATHER WM. F. ROACH, Pastor
St. Mary's Catholic Church
Texas City, Texas

This Space Paid for by Citizens of Texas City

As he walked back onto Sixth Street, Roach knew his letter would be clipped and marked "for immediate delivery" to the members of that "steel band." It would be sent to the St. Louis office of Monsanto president Queeny; to the Fort Worth office of eccentric Texas billionaire oilman Sid Richardson, the richest man west of the Mississippi; to the New York offices of the Rockefeller family.

He also knew that, quietly, Mayor Curtis Trahan agreed with everything he had written. The two men had become unlikely friends, allies. Roach was from blue-blooded society circles in Philadelphia and had somehow washed ashore in Texas—and then reinvented himself as a combative cleric, a champion of the workingman. Trahan was a towering, taciturn Texan who had only a high school education and spent years doing the nastiest blue-collar jobs in the oil refineries. Then he had come home from World War II as a profoundly wounded hero who at first seemed to be the perfect, malleable man for the window-dressing position of mayor in a small town run by giant corporations.

Roach knew that Trahan would be fielding angry phone calls about the damning letter.

Roach also knew that, as always, there would be wildly varying opinions: *Bill Roach is a saint. A socialist. A communist. A nigger lover.*

One thing had always been clear: Before he had even unpacked his one piece of luggage at his modest home on Third Avenue North, Roach was famous along the Gulf Coast. His work, his inclinations, had been followed in religious circles for almost a decade. He was widely recognized as someone who was assigned to the most difficult cities—the places where poverty, prostitution, and a plague of social ills had a stranglehold.

"A saint is coming your way next week. His name is William Francis Roach," the awestruck priest preceding him at St. Mary's told the congregation.

Trahan had watched in amazement as Roach elected himself the town conscience—the finger-wagging embodiment of the maxim about comforting the afflicted and afflicting the comfortable. For Roach, Texas City was a wide-open laboratory, a booming, sweat-stained, billion-dollar workshop where he could dig into the internal mechanics of an entire American town.

And up until two weeks ago, up until he began to be eaten alive by his own gruesome visions, Roach had felt he was in control of some things—that he had found a way to harness the giant forces that had made Texas City unlike any other place in America. Roach once told his brother that he was at the absolutely right region of the United States at the absolutely right time.

It was also something he once confessed to Curtis Trahan: Texas City was the exact center of everything that really mattered. The biggest tugs, the biggest themes were all boiling over in Texas City.

Power. Race. Money. The Cold War.

Why the hell would you want to be anywhere else?

Now Roach felt as if there was no escape from Texas City.

THE SAME DAY his open letter ran in the local newspaper, *Time* magazine featured a breathless report on the dazzling revolution unfolding in Texas City. Its April 8 edition said that the region was ground zero for a new, profound era that would forever change America and the world: *"The burgeoning Age of Chemistry"* was at hand, and *"Texas was well on its way to becoming the chemical capital of the world."*

Time noted that 145 new chemical plants had been built along the Texas coast in the last seven years. The companies were spending $1.5 billion to build even more factories, hire thousands more workers, and ratchet up experiments and production.

"The complex equipment of modern manufacturing has spread its twist-

ing tentacles all over the once wide-open spaces. . . . But Texans by and large are for it because it has given them something new to brag about and it will make them richer. They do not mind that 90 percent of the new investments have so far been made by outsiders."

The war had just ended with startling proof that American scientists could go into the raw essence of energy, of life, and emerge with the ultimate weapons. Edward Teller, Robert Oppenheimer, Enrico Fermi, Albert Einstein, and the other men who lent their theories to the atomic program were the embodiments of America's open arms and invincibility. Recriminations about what they had unleashed in Nagasaki and Hiroshima hadn't fully taken hold. America had placed great faith in its scientists. And war, mankind's greatest dilemma, had been solved.

Other revolutions were unfolding in the laboratory:

American chemists were unleashing 2,4-dichlorophenoxyacetic acid—and they had given it the ability to choose between life and death. It was patented as the first selective plant killer, bringing death to weeds but, amazingly, leaving adjacent crops unharmed. Chemist Dorothy Hodgkin was solving penicillin's deceptive structure; it was going on the market commercially and would save millions of lives. American chemist Willard Libby was perfecting work with carbon-14 dating, the breakthrough method that would allow researchers to determine the age of buildings, treasures, parchments, and clothing. It would allow mankind finally to chart its own history.

Tetraethyl pyrophosphate, discovered by secret agents inside Nazi chemical warfare labs, had been converted by American chemists into the most powerful insecticide in the world. Chemists urged the city of Grand Rapids to introduce a "miracle" element called fluoride into its drinking water. Chemists were testifying to the safety of selling countless loaves of bread "enriched" with items from the laboratory. Meanwhile, millions of tons of high-powered nitrogen compounds, including ammonium nitrate, would double, triple, and quadruple farm yields . . . and make things grow where they had never grown before. Simply put, they would make them grow where the earth's soil had been so drained of nutrients, so ravaged by overuse and world war, that the land had often been left for dead.

The chemists even extended their work to the ordinary American kitchen: a Dupont chemist named Earl Tupper was busy designing seemingly indestructible bowls that he wanted to call Tupperware. And now, children around the country were reading *Popular Science* adventure stories about their white-frocked laboratory heroes. Christmas packages contained beginner's chemistry sets. In the pages of *Time* and other magazines, the Age of Chemistry was being promoted, packaged, and endorsed. The ads—for asbestos, nickel, Lustron, Skylac, and a hundred other minerals, chemicals, and petrochemicals— were everywhere:

"Patients of Zulu Witch Doctors sometimes have to swallow a dose of asbestos . . . for this medicine man carries powdered asbestos among his pharmaceuticals! It has as many uses as a building has surfaces, whether for farm buildings, industrial sheathing, or house remodeling jobs. It never needs painting, fire can't hurt it, time only toughens it. Best yet, its cost is moderate and maintenance is practically nil."

It was enlightened capitalism, unshakably patriotic proof that America had the commitment to tame nature. "Far-sighted men" were brewing chemical creations that were almost heroic, and in Texas City, Monsanto was dedicated to turning petroleum and petrochemicals into a new product called plastic. Enchantingly, Monsanto renamed it "Lustron":

"Some far-sighted men took an idea to a molder, produced it in Monsanto Lustron . . . they capitalized on . . . qualities that no other material combines so advantageously as Monsanto Lustron."

The endless advertisements insisted that a problem posed was a problem solved. Petrochemical solutions would be tailored to each emergency.

"The Sky's No Limit . . . Skylac, the answer to the challenge of the aviation industry, is only one of many 'custom-made' answers Monsanto has for the . . . problems of all industry."

One magazine ad showed a bride in a stunning, billowy gown, depicting America happily wedded to the Age of Chemistry:

"The bride wore coal, air, and water . . . and your 'unseen friend' was there. Of such stuff dream-dresses are made. Yes, made by the chemist, who can transform commonplace materials like coal, air, and water into glamorous materials."

In Washington, President Harry Truman was well aware of the scientific advances—and he came to the realization that they could be used to win the dangerous new war waged by the Soviet Union and the United States for the hearts and minds of millions of people around the world.

Three weeks earlier, Truman called an emergency joint session of Congress:

> *Mr. Speaker, Members of the Congress of the United States:*
>
> *The gravity of the situation which confronts the world today necessitates my appearance before a joint session of Congress.*
>
> *The peoples of a number of countries of the world have recently had totalitarian regimes forced upon them against their will. . . .*
>
> *We must take immediate and resolute action. . . . This is a serious course upon which we embark.*
>
> *The seeds of totalitarian regimes are nurtured by misery and want. They spread and grow in the evil soil of poverty and strife.*
>
> *If we falter in our leadership, we may endanger the peace of the world—and we shall surely endanger the welfare of our own nation.*

Truman wanted crops suddenly to appear in those once-dead, battle-scarred fields in Germany, France, and Italy. Truman wanted millions of pounds of a life-giving chemical compound called ammonium nitrate to be churned out in government plants.

Texas City would handle it. Texas City would ship it out. Texas City would help save the world.

If only he could sleep.

Running a brush over his black hair, Roach stared at his face. There were mottled purple hollows under his eyes. They seemed painted onto his sallow skin, which was covered in a sheen of sweat. Roach had never had lingering health problems—one time, when he had diabetes, he decided to skip trips to the doctor and essentially will himself to be better. He no doubt knew people were studying him, and that would probably only make it worse. The noises from outside seemed amplified. The squawk from the dull-gray gulls, the sound of the foghorns, the drone from an oil company airplane, the restless chatter from the children heading to the nearby elementary school. It was loud, too loud for him to be able to pray.

What the hell are people thinking about me?

During his Sunday sermon a week ago, a heaviness pressed down on his congregation at St. Mary's that was far beyond the usual sacred interludes. Bernice Smith and the other parishioners present that Sunday would talk about it for decades after Roach's stunning, mysterious death. Roach began:

"Easter is coming. But we need to lead better lives. We need to change our ways. Texas City is a wicked place, a very wicked place. If we don't change our ways, blood is going to run in our streets in a very short period of time."

That unsettling Sunday, a quivering, pale, sweating Roach stepped away from the pulpit with the eyes of his rapt, faithful parishioners trailing him. The air was thick with the smell of chalky incense and melted wax. Not a soul in the church moved. The usually creaky wooden pews, warped from the constant coastal humidity, were silent.

Roach had decided not to announce that he, too, was going to die.

He didn't know what he would possibly say in his next sermon.

He didn't know if he would be alive to deliver it.

Hell, he didn't know if anyone would be alive to hear it.

After his parishioners gingerly left the church, he closed the doors. In the darkness, alone, he would probably light another cigarette.

The Ship

THE GRANDCAMP *should have shoved out of Houston already, but immigration officials and FBI agents are still crawling all over the long, rotting waterfront. You can see them in their late-model cars, pulling into the dingy parking lots of the low-rent Greek and Chinese diners where passed-out sailors sometimes have to be hauled out, their heels scraping across the soiled tiles. You can see them with their fedoras pushed back on their heads, huddling and conferring at the restaurant tables, reviewing their lists and then retreating to a telephone booth.*

On the docks, the government agents pull the French captain aside to grill him about the fact that some of his sailors have Communist Party papers in their passports. Stoker Pierre Andre is one of them. Now, everyone will have to wait. Everyone will have to watch the maddening swirl of petrochemical smoke as it melts into that ash-colored sky.

It is as if Andre and the other Frenchmen are being tied down by an insistent web of history—as if they can't escape the ghosts lingering on the sluggish, shrouded bayous, channels, bays, and veins of brackish water: this tattered ribbon of the Gulf Coast was once overrun by pirates—by the hardened compatriots and bastard offspring of their more infamous predecessor, the French pirate Jean Lafitte. The buccaneer had built his own city on nearby Galveston Island, crafted his own constitution, and then oversaw it all from a wretchedly splendid house called Maison Rouge. Lafitte held lavish parties and built slave pens to house the purloined Africans and Caribbeans he had plundered from Spanish galleons. Plantation owners in Mississippi knew enough to make the journey to the Texas Gulf Coast to buy their slaves directly from Lafitte—and when they arrived they saw bodies hanging from gallows, prostitutes tripping through the thick mud and scummy ponds, former U.S. Navy heroes arriving and enlisting with Lafitte.

When government warships finally ordered Lafitte to abandon his city of a thousand whores, pirates, and swindlers, he sailed into the Gulf of Mexico and was never seen again. A handful of his pirates risked staying behind, and they were joined by a ragtag ensemble of rootless souls from around the world. They carved out a dangerous life inside the clouds of mosquitoes and on land that was the color and consistency of bruised fruit. They built homes, docks, and piers over the marshes and resigned themselves to the fact that, sooner or later, most of what they had built would groan and crumble in the next marauding storm.

It was, until oil and chemicals changed everything, a place favored only by the reckless, the untamed, the wanderers. It was, in many ways, one of the most forgettable places in America. And even now Capt. Charles de Guillebon, Pierre Andre, and the sailors aboard the Grandcamp can see very few reasons for lingering. They want to follow Lafitte, to leave Texas, to disappear into the Gulf of Mexico.

De Guillebon watches the U.S. agents confronting his crew. He knows the Americans are stalking the Communists these days.

The damned all-enveloping fog is clogging the ship channels at the northern end of Galveston Bay. Veteran Houston sailors have told him to be cautious about making his way down the twisting basin. They've been saying it is one of the most dangerous in America, a place so narrow and shallow that a man had to be on the watch for any trouble rising out of the night. The captain doesn't want to bull-rush his way into the Texas City waters. He'll wait out the interrogations from the FBI as well as the choking fog.

Texas City isn't going anywhere.

The Company

April 9, 1947

THE BAROMETER has been seesaw-
ing, and there is a buzzy, electrical taste to the air. It should rain good
this weekend. Out beyond the steel-and-iron petrochemical complex,
out toward Bolivar Peninsula, a small squall has formed, and it looks
like an enormous gray willow tree twitching and stretching from the
horizon to the stack of black clouds above. The small shrimp boats that
have ventured into the shallow waters of the Gulf are arcing back into
the bay.

And, right now, just thinking about William Francis Roach is making Curtis Trahan shake his head in disbelief.

The biggest decision in Texas City history will take place in about
twenty-four hours. Curtis Trahan and the four city commissioners
have a meeting on the second floor of City Hall to vote on whether or
not finally to annex the entire industrial zone. If it passes, Texas City
will be twice its present size, and millions of dollars in tax revenues
should begin steering into the city—or, in a worst-case scenario, the
billion-dollar collection of oil companies and chemical plants will
decide to move away. Maybe, if the companies are annexed, Texas City
can avoid what it had to do last year—sell off its only fireboat because
there wasn't any money to pay for maintenance.

Curtis knows Bill Roach will be there.

He doesn't have to wonder where Roach stands on the issue. He can damn well hear Roach's ballsy voice when he picks up the *Texas City Sun* and rereads Roach's open letter.

The big war has been over for only eighteen months. Exactly half the people in Texas City have either worked in a war plant or served overseas—some have done both. Hitler and the Nazis are still the faces of evil—and Bill Roach's giant public letter says that the people who provided Texas City's lifeblood are Nazis. His letter said the city was run by "selfish individuals" who had formed a "steel band" to choke off innocent people—just as Hitler and the Nazis had done in Europe.

> *"Hitler. Nazis. Steel band. Selfish individuals."*

Monsanto. Union Carbide. Amoco. Humble . . . all of them.

All morning, Trahan has been meeting people quoting the rest of the damned thing:

> *. . . trapped within the "steel band" . . . these poor people were also created to the image and likeness of God. They too have an innate right to share God's good earth, and the right to live as human beings, to own decent homes and property. They too have worked, slaved, fought, and suffered that our America might be great. Again, we are brought back to the center of our vicious circle . . . most cities have municipal conveniences (necessities) as well as luxuries. Texas City has neither! It is not the wish nor the desire of the writer, or of any citizen, to "bleed" industries . . . but rather to ask that they support, in a nominal way, the city in which their employees live. . . .*

Trahan has heard this from Roach before.

Roach, his black robes flapping, is usually in a heady rush, ramming from one end of town to the other. He's always pacing Texas Avenue, the smokestacks and towers in the background, and talking to kids like Forrest Walker Jr., kids whose fathers moved their families to town and are just now punching the clock as boilermakers, chemists, and engineers inside the Monsanto plant that sits five hundred feet away from Pier O. When a Negro longshoreman like Ceary Johnson decides to do the unthinkable—to vote to lead his gang members on a strike for equal pay—Roach is on Johnson's doorstep in The Bottom:

I'm going to walk with you.

And, when the strike finally uncorks, Roach is on the waterfront, locking arms with the edgy dockworkers as they try to block the path between the ships and the nearby railroad terminal. He offers sand-wiches and then offers himself as a shield against the strike-breaking Texas Rangers with their long rifles pointed out the window.

Back at the regular nighttime City Hall hearings, Roach is always in the first row.

He's waiting his turn under a brown ceiling fan that's barely moving the stale air. He is bobbing impatiently in his wooden folding seat, looking for allies: maybe Trahan, the army private whose left leg had been almost blasted off in a Belgian forest—and who somehow didn't die and who somehow got himself elected mayor despite his populist inclinations.

It took some time, but Roach no longer believes that the physically imposing but surprisingly soft-spoken Trahan is going to be a patsy for the Rockefellers, the Bushes, Amoco, Humble Oil, Monsanto, Union Carbide, the Santa Fe Railway, and the other out-of-town giants. At the latest City Hall meetings, Roach has been stunned. Trahan has been quietly standing his ground, especially on the revolutionary notion that Texas City needs to cure the wretched, piss-poor living conditions for the low-paid union workers and the families in The Bottom and El Barrio.

Roach, for a while, had been completely on his own.

He felt like a martyr.

ONE DAY, ROACH invited longshoreman Ceary Johnson to come to the priest's cottage at the Third Avenue Villas—to eat homemade tamales he bought from a Mexican lady down in Spanish Town, to talk about unions, to discuss Roach's blueprint for opening a bank for blacks.

As Johnson steps to the front door, he can't believe he has actually crossed that palpable Texas Avenue checkpoint separating the black and Mexican neighborhoods from the white side of town. Sitting across from Roach that day, he can't believe that a white man is asking his opinion on the fate and future of Texas City.

There are, after all, black men and women living on Second Avenue who remember that decades ago the locals staged minstrel shows to raise funding to lay down the foundation for St. Mary's. And down in The Bottom, down in Nigger Town, the men playing dominoes at Willie Green's Café still talk about the Negro who was crazy enough to cross Texas Avenue. He had a new Buick. He had gone north, up to Sugar Hill, into the white neighborhood. The Texas City police chief chased him down, in the shadow of the looming refineries, and said the nigger was reaching for something in the backseat of his car.

He had to be shot.

Roach is a nigger lover. Read his letter.

Maybe he'll be shot, too.

THIS MORNING, a low-slung flatbed truck is crunching over the pock-marked streets and through the little patches of scummy water bubbling with mosquito larvae. Rocking in the truck bed, the six members of Ceary Johnson's longshoremen's gang are half awake but hopeful, wondering if they'll snag one of the disappearing jobs hauling dry cargo instead of the six million barrels of oil that's constantly being shipped, tanked, piped, and pumped from the legendary Texas oil fields to Texas City—so the refineries and Monsanto and Union Carbide can heat it, crack it, reduce it, and mix it into the plastics, varnish, sealer, additives, rubber, heating oil, lamp oil, automotive gasoline, and aviation fuel.

Just off Second Avenue in The Bottom, Johnson's truck passes a black man with a mule train. A tall, weary man named Carl Templeton has been up since midnight collecting human waste left in buckets in the alleys behind the shacks. There is no running water for most of the homes in The Bottom. Rags and smooth pieces of driftwood are stuffed into the uneven window frames to keep out the stench of the feces and the petrochemicals.

As Johnson's truck moves out of the neighborhood, heading to the industrial zone and the docks, it passes something else.

Inside both El Barrio and The Bottom—and loosely bound to each other—are the half-dozen brothels and beer joints for the foreign crews, the wanderers, and the unfaithful who are seeking some kind of diversion the minute they exit the refinery gates.

The joints have been there for years, next door to where the Garcia family lives, where the Johnsons live, where people go to church at flimsy-looking Our Lady of the Snows—the improbably named community anchor that's a holdover from the days when Mexicans couldn't worship at St. Mary's. There is The Sailor's Retreat, a two-story building with a stained bar downstairs and a flophouse upstairs. There is Jeanette's Place and Tillie's, which actually looks like someone's modest two-story home, until you spy the silky women's undergarments fluttering on the clothesline that Tillie Davis set up in the backyard. There is The Stumble Inn, smelling like stale beer, tobacco, and sweat. There is The Ship Ahoy. There is a place some people call The Beacon, crowned with a lighthouselike lantern that slices through the petrochemical puffs and summons the restless from anywhere along this stretch of the Texas coast.

There is a joke about the lantern:

It was the best damned lighthouse around. You can see it all the way to Havana and Key West, and it saved many a sailor's life.

Roach and Trahan swear the whorehouses will go.

AT THE DOCKS, Ceary Johnson and his longshoremen wait some more . . . and wait . . . until the men from what many people call The Company bustle out to meet them. Nothing significant has ever arrived in Texas City without the permission of The Company.

This stretch of the Gulf of Mexico was developed in the late 1800s by Minnesota investors with deep pockets and wild ambitions. Their plan was to do anything to sculpt the old pirate's outpost into a brawny Lone Star version of New York City. In the most self-congratulatory state in the nation, it was going to be the state's centerpiece, easily the most important port in Texas. It was going to be as powerful as any other city—be it New York City or Oklahoma City—that dominates the state after which it is named.

Picking the name for their new municipality was easy braggadocio: *Texas City*.

A pair of quasi-official, all-powerful firms—the Texas City Terminal Company and The Mainland Company—were created. Most people in Texas City recognize the fact that the same business interests run the city, and many simply refer to both firms as The Company. It has control of almost everything and seemingly unlimited resources and political connections. The Light Company, the Water Company, the Port of Texas City, and other "companies" are just subsidiaries serving a common goal. To oversee it all, Col. Henry Moore—a personal friend of Gen. John Pershing and one of the world's logistical geniuses when it came to moving massive amounts of men and supplies (Pershing put him in charge of U.S. troop-and-supply transports during World War I)—was brought to Texas City.

In time, the colonel oversaw the sale of The Company to a consortium of the biggest railroads in America, and Texas City remained the largest privately controlled port in the state.

Moore and the men who ran The Company embodied a certain way of life in Texas City.

They lived where blacks and Mexicans were never allowed to go—unless they were servants. The Ku Klux Klan quietly enforced the segregation, underscoring the opinion that some people are only fit to live down where the ground is so close to sea level that it is like walking through gumbo . . . down where the foreign-speaking sailors find the whorehouses and the "chock houses" that sell sealed jars of "chock" or homemade liquor.

The KKK was there.

"Let's say that they were thought to be the monitors of our morals. They also got into politics, too," said the colonel's closest aide, Thomas Bynum.

And in the end, Texas City was three cities in one:

There was the waterfront itself and all that it implied—crews from Mexican freighters splayed out on the docks at night, smoking cigarettes and roasting some fresh meat; dockworkers, longshoremen, and refinery men leaning against the pilings and carving hunks of cheese and apples for lunch; the black man with a guitar slung over his shoulder who had snared a free train ride to Texas City and was playing music for tips; the bloodstained fishermen cutting up their redfish and shrimp.

There was that zone of poor neighborhoods—El Barrio for Mexicans and The Bottom for blacks—crammed close to the docks and reeking of fish, oil, and chemicals.

Finally, there were the uptown neighborhoods, crossing north of Texas Avenue, where the homes had verandas and even a second-story widow's walk—where the families of the sea captains could stare out toward the Gulf of Mexico and count the days until the big ships return to port. Some of the homes had tin roofs and Victorian gingerbread designs. Some were framed by sentinels of palm trees. They are set back from the street and the walk to the wide wooden porches; the hanging flower baskets and the gurgling fountains reinforce their splendor.

The colonel's house was there. So was his elderly black servant, the one he had been given by the governor of Texas.

"The governor had signed him over to the colonel," remembered Bynum's wife. "A little old dried-up colored fellow. The governor knew the colonel, and he let him have him. He'd signed him over to the colonel, as long as he behaved and as long as the colonel was satisfied with him, he could stay there."

In The Bottom, in Nigger Town, people said the colonel once had his own slave.

In The Bottom, there were people who were convinced that God would one day punish Texas City for some things in its past. There were people who thought God was an avenger.

Now, as CEARY JOHNSON and his men linger at the north slip, two determined-looking Company men are in conference under the spreading haze: There is the president of The Company, Mike Mikeska, whose voice sounds like rusty nails rattling in a can; there is his cigar-puffing, fedora-wearing vice president, Walter "Swede" Sandberg. They are clutching the bills of lading from the Santa Fe Railway, conferring with shipping agents, haggling with the union representatives, examining cargo inventories from The Company's warehouse foremen, and directing the movement of gangs, crews, foremen, and tugboat captains.

Mikeska has studied under Moore and the millionaire investors who have come down from the upper Midwest and willed Texas City into an improbable existence. He has been with The Company for three decades. When Amoco finally arrived in the 1930s and built its huge refinery, he decided it was like Jesus coming to save the city.

"A savior," Mikeska likes to say.

And now, he'll be damned if Curtis Trahan or that strutting rooster Bill Roach will annex and tax the big oil companies on the waterfront. Mikeska has already decided he'll be at that big annexation-taxation meeting at City Hall. He'll either delay the vote or have the whole goddamn thing thrown the hell out.

Mikeska meets regularly with the refinery representatives; he meets with them whenever they summon him. He is the intermediary, the ambassador, between Texas City and the national oil and petrochemical companies.

The colonel had promoted Mikeska through the ranks, grooming him as his eventual replacement. And the eager Mikeska watched as Moore relentlessly sold Texas City as a place where business and the military could damn well do anything they wanted. The wetlands and marshes were easy to carve, and the oil could be piped, shipped, and processed there—converted into a rainbow of petrochemicals and then easily shipped out again.

Moore always reminded the companies that the best land for business in Texas City was in the parcels close to the waterfront, down by the docks and the railroad terminal. It all happened to be in a zone just outside the city's original town limits—a zone miraculously free from taxation.

Now, Texas City was one of the world's most important tidewater terminals. It was receiving rivers of crude pouring out of the East Texas fields and the newer, gurgling West Texas fields—and it was processing and shipping millions of gallons of refined Texas black gold to New York and far beyond.

When World War II exploded, Texas City was ready.

The problems posed by the U.S. generals—supplying Allied troops with tires for military transport trucks, medical tubing, airplane wheels, engine lubricants, airplane fuel, Jeep parts, pontoons, cans, radiators, printing presses, chains, buttons, pins, engines, ammunition casing, and hundreds of other essential items—was solved by Texas City.

The largest tin smelter in the world was built in Texas City. The land was given to the Texas Tin Corporation for a dollar—and it happened to be in that zone near the waterfront free from any city taxation. The gigantic plant went on line in a matter of months, handling forty-five thousand tons of tin a year, enough to supply all the military and civilian needs of every Allied nation.

The tin bars it shipped out around the world were stamped with the image of a Texas longhorn.

MIKESKA KNOWS THE big French freighter is due in any time soon from Houston. There are twenty-six hundred tons of ammonium nitrate sitting in Warehouse O.

The endless stacks of fertilizer bags in The Company warehouses have been rumbling into Texas City over the last few weeks from three military plants in Nebraska and Iowa. Longshoremen moved them off the trains and into storage alongside the docks. Now several gangs will get to work using winches and cranes to hoist the hundred-pound bags onto the freighters. From there, the gangs will descend belowdecks, into the dark holds, to stack the bags and fill up every available inch.

Pier O, on the north slip, is where the dry cargo work is done by Ceary Johnson, Julio Luna, and the others. There are usually fat sacks of peanuts from South Texas. Rice from Crowley, Louisiana. Still-moist wood from Costa Rica. Burlap bags of ore from the Bolivian tin mines. Coiled twine, made from agave hemp in the Mexican interior, that will be shipped up the Missouri River and delivered to prisons so that inmates can fashion it into rope.

There are also masses of cotton that only a damned good "cotton header" can handle. Bill Roach has seen them work. The cotton headers are muscled, sinewy artists—honored in any waterfront bar for being able to sprint with wooden carts holding piano-sized clumps of cotton, aiming the stack straight from a Santa Fe train and into the pitch-black hold of a Liberty ship.

Driving by the waterfront, Roach skids to a halt, puts an elbow on the ledge of his car window, and marvels at the cotton headers. Sometimes Trahan is already there, in awe of the workers, their rhythm, the buzz, the hum from everything clicking and coalescing on the waterfront. There are saints, sinners, drinkers, smokers, teetotalers, Holy Rollers, and plenty of men who picked up wickedly saucy tattoos from their days serving overseas.

Roach blows a cumulus of smoke out the window of his Ford and takes one last look at the gangs of men sweating, cursing, and putting their shoulders into it on the Texas City waterfront.

They look like ghosts and they look like dead men.

They look as if they're carrying the entire weight of Texas City on their shoulders.

He was sure of one thing about the doomed souls in Texas City: They all showed up for goddamn hard work and down here, down on the unforgiving Gulf of Mexico, that act alone predestined every single one of them for a chance at getting to heaven.

And with the smoke rising and curling around his face, Roach prayed that the people in Texas City would cross the bridge to heaven much later than sooner.

The Ship

PIERRE ANDRE *and his shipmates are racing through final-hour preparations, secreting away stashes of cigarettes, chocolate, and the things with which sailors reward themselves when they're weeks away from anything that's familiar. With any luck, they'll be home soon. The word from the deck is that the captain thinks there's an opening, a window through the fog.*

Andre and his mates are in and out of the thudding engine room, with the familiar echo from the three-cylinder reciprocating steam motor fired by twin oil-burning boilers. Andre and the crew have worked like dogs to keep the ship running smoothly. Like intinerant sailors everywhere, they had built a certain sort of grudging affection for the Grandcamp. *It was a warhorse, just like the sailors. It was a journeyman freighter, manned by journeymen sailors who had signed on for one more run across the oceans.*

And like some of the sailors, the Grandcamp *was lucky it was still at sea. It had been built five years earlier in Los Angeles—christened the* Benjamin R. Curtis *and hammered together in seventy days as part of a fevered American wartime program to crank out ships that could deliver troops, guns, ammunition, tanks, and airplanes anywhere in the world. It was a World War II Liberty vessel—one of 2,751 freighters that were built, one after another, in eighteen different shipyards on either coast.*

"Ugly ducklings," trumpeted President Franklin Delano Roosevelt when

he first saw them—but every single Liberty ship was a blocky testimony to America's fierce skill at racing to war. Some of the Liberty ships were born in under a month. The Robert E. Peary was done in four and a half days.

The ships were thrown together with the same blueprint: The 250,000 parts for each of the ships were prefabricated in thirteen states in fifteen different plants. Multiton chunks were loaded onto railroad flatcars, hauled to the nearest shipyard, and assembled by men and women working around-the-clock shifts. Each was given the name of a prominent but deceased American: Polar explorer Peary. Dead-eye sheriff Wyatt Earp. Supreme Court justice Benjamin R. Curtis.

When the Curtis was launched by the California Shipbuilding Co. in Los Angeles in November 1942, it looked like all the others: 441 feet long. She cruised at eleven knots. The holds could handle nine thousand tons of supplies. There were small antiaircraft guns, almost as an afterthought, on deck. It was another blue-collar truck on the ocean, resolutely crisscrossing the Atlantic and the Pacific, making up for a lack of grace with squatty muscle.

The Curtis served in the Pacific theater, making endless runs, and when the fighting ended, it was mothballed at Hog Island in Philadelphia. A year later, Washington ordered some relatively seaworthy Liberty ships to be made available to the European Allies. The Cold War was dawning. The ravaged continent needed to be restored and protected from the Soviet Union, with the massive infusion of coal, cotton, and fertilizer that a Liberty ship could deliver.

The Curtis was assigned to France, and officials from the French Line— home to historically sleek thoroughbred ocean liners like the Normandie— issued orders for forty seasoned sailors, including Pierre Andre, to go and claim the decaying cargo ship in the United States. Each of them sailed to America and converged on Philadelphia. From there, they steered the salvaged vessel to New York. They chipped away at the rust, primed the engines, and lashed down gear as a hedge against the towering, icy seas.

Ship painters lowered themselves down the hull and put S.S. Grandcamp over the name of the U.S. Supreme Court justice. They chose the name in honor of a French harbor town that had been occupied and heavily defended by the Germans for four years until it was liberated by endless, crippling Allied attacks from navy destroyers, from Rangers, and from the men of the 116th Infantry Regiment of the 29th Division.

Now the French sailors will be helping feed their families and a million more with an ex-American ship filled with life-sustaining American goods—including some magical chemicals made in the same bomb-production plants that once helped to destroy the European strongholds of the Axis powers.

And now the rank smell that moves like a grimy hand along the under-belly of the Texas Gulf Coast is seeping into the lower berths of the Grand-camp. *The journey has been interminable, but, finally, Andre has a sense that the end is near.*

He's not alone.

Puffing on cigarettes, the smell of tobacco thankfully staving off the dizzying aroma of the petrochemicals, the sailors are hunkered under weak harbor lights and writing their last letters to the women and children they have left behind.

The Priest

ALL WEEK, the parishioners at St. Mary's have been huddling to whisper about Roach's alarming, unembarrassed way of mingling with the spirits. After the morning services, nervous people gather in clumps and trade alarming stories:

There is Roach bolting down the aisles of his church. There he is brushing past the dull-ivory-colored statue of Mary and booming out: *"I can't talk right now, but I'll talk to you later."*

He is addressing his comments directly to the statue. Waving at Mary. Just the way he would talk and wave to someone standing on Texas Avenue or heading into Lucus's for a deluxe hamburger. Then he is caroming out the door, reaching for a match, and disappearing in another blast of Old Gold smoke.

There are other stories: After earnest doctors informed him that his diabetes was going to lead to his left leg being amputated—and that he would probably also fall victim to a premature death—he sank to his knees and prayed.

Save me, Mother Mary.

Then he went to a movie, ate a pound of candy, drank a milkshake, and returned to his doctors for a round of glucose tolerance tests. The physicians were stunned. The new tests showed he was no longer diabetic. Roach never talked about it, except with his closest friends, but a handful of them were already drafting the proper paperwork to send to officials at the Vatican. It seemed like a verifiable miracle, the kind of event that was a prelude toward sainthood.

There is one more incredible rumor gingerly being passed along with urgent request that it be kept extremely confidential: Roach has become an adherent, a follower, of a nun in Detroit who bears the bloody signs of the stigmata on her hands. He has gone, with his brother, John, to visit the woman, and she has shared visions and prophecies with the duo. She has talked about seeing a house along a railroad track in Texas, maybe a place where the Roach brothers can build a secluded sanctuary.

She has talked about seeing something awful. She was the one who told him what would happen in Texas. She was the one who put the vision in his head.

"There will be blood flowing in the streets of Texas City," the stigmatic nun told Roach.

PEOPLE WHO KNEW Bill Roach growing up would never have placed him in the gritty, ball-breaking underbelly of America—let alone being trapped by nightmares.

Sometimes Roach thought his being steered to Texas City had everything to do with his mother:

She had been a well-off Philadelphia matron wrestling with a troubled pregnancy. She refused medical treatment, and doctors warned that she could die. A devout Catholic, she insisted on being brought to the hospital chapel.

Dear Mary, I beg you to care for my children. I dedicate them to you.

Later that day, she gave birth to her twins. She died shortly afterward.

Their father, whom they most resembled, ran profitable service stations and shuttled the boys from boarding schools to aunts, uncles, and grandparents. Johnny and Bill moved on the fringes of educated, high-society Philadelphia circles. They took up competitive rowing and mingled at soigné parties attended by a toddler named Grace Kelly, the future princess of Monaco. They were fey, whip smart, and hell-bent on keeping their days perfectly dangerous. The twins formed a neighborhood gang, a band of six bored and well-off young men who were prone to stealing bottles of holy wine from a local church.

Their father was worried. So were most people in the family.

The bond between the prodigal sons went well beyond the usual ties that bind. They drifted through the same jobs in the greater Philadelphia area: driving trucks, working at fruit stands, civil engineering. As they entered their mid-twenties, friends and relatives finally saw them beginning to shake loose their dicier tendencies. But, still, few people were prepared for the latest bombshell:

Bill Roach had slammed his car to a halt in front of a small church in the Pennsylvania countryside. In a spurt of self-reflection, he suddenly wanted to unburden himself, abandon some guilt, by going to confession. The country priest concluded that he was being visited by an idle rich wastrel in search of direction. He posed a question:

Why not contemplate the priesthood?

Not long after, without knowing his brother had been there, John stopped at the same church. He had the same discussion with the same priest, and he received the same advice.

Bill Roach started going to mass every day. His brother followed suit. In 1936, they announced that they wanted no part of the family business. Instead the Roach boys were going to thunder out on a mission to marry religion with righteous indignation. The South seemed like fertile ground, like a perpetual adventure in a troubled place that could use perpetual healing. The brothers loaded a few things into their Ford.

The plan was to aim for Arkansas and petition bishops for admission to the priesthood. Before the twins left they asked a nun, a teacher at their high school, to write a letter of recommendation. On the road, with his brother behind the wheel, Bill unsealed the letter.

He read it aloud:

"I cannot recommend these Roach twins for the seminary. As far as I am concerned they have no religious or priestly vocation. I would recommend anyone else but the Roach boys."

Bill stared at the stinging letter and then began to rip it into small pieces. Laughing, he threw the paper bits out the window and watched them flutter away.

When they reached their destination, a seminary in Little Rock that was atop a hill bordering the Arkansas River, Johnny said: "The Sister had given us a letter of recommendation, but we can't find it in the car."

The Roach boys were admitted anyway. They were convinced it was going to be good. They began studying in earnest for the priesthood, each of them thinking he was going to change the world. They stayed for a year, until they were told that the seminary was running low on funds. They decided to try Texas, where they assumed the sins would be even greater, where the Gulf Coast was exploding with dreamers riding the giant oil wave.

When the Roach brothers left Little Rock, their friends saw some huge graffiti painted two stories high on the brick side of the administration building. It was a single word:

Roach.

IN LATE '37, they rode the Ford over the causeway that spanned Galveston Bay and linked the Texas mainland with Galveston Island. They raced down Broadway—Galveston's mansion-lined artery—and pulled to the front of the Bishop's Palace. It was a frothy, spiraling architectural concoction of Renaissance and Victorian styles. The intimidating place had been finished in 1893, somehow survived the most devastating hurricane in American history when it blasted through in 1900, and now glowed with silver and onyx mantels as well as yards of rosewood, satinwood, and white mahogany.

John, better at seducing the bureaucrats, told his brother he'd do the talking. Bill restrained himself and said he'd stay in the car.

Ushered inside, John waited under high ceilings for his audience with the famous Bishop C. E. Byrne. The church elder listened politely and said there weren't any openings. From the second story, he looked down and spied the beat-up Ford.

"You mean you came all the way here in that?" the bishop said.

John replied: "There's one more just like me in the car."

Byrne chuckled. He made arrangements for their admission to a nearby seminary. The Roach boys were ordained the same day in 1939, they celebrated their first masses within hours of each other, and they each set about breaking most of the rules, unspoken and spoken, for novice priests.

Simply put, no one in Gulf Coast religious circles had known anyone like the calculating, combative Roach boys. They were the primogenitors of their own southern brand of liberation theology—the fiery belief that religious leaders had a moral obligation to get mixed up in real-world problems. John settled in Houston. He confronted the Catholic hierarchy, clamored for the church to build inner-city hospitals, and took trips to New York City to study radical notions of social work with labor leaders and activists.

Bill hit the rural trail, going door to door in one suspicious Texas town after another, constructing sanctuaries in places where Catholics were viewed as idol-worshiping papists. With church elders keeping a wary eye, he was Johnny Appleseed in a collar, scaring up the money and the workers to build one more church and community center— and bracing himself for organized resistance from Ku Klux Klan vigilantes who hadn't encountered fast-talking Yankees from Philadelphia. Especially ones who immediately opened the church doors to black sharecroppers, Mexican fruit pickers, and anyone else who wanted to come in.

Shirt off, cigarette dangling from his mouth, Roach climbed scaffolding, hammered nails, and seemed to be up earlier and later than anyone else. At a half-dozen parishes in the outbacks of sweaty, barebones Texas, Roach was opening churches, community centers, medical facilities, and schools. He covered an area of 2,791 square miles. He also served as the chaplain at nearby Fort Hood, where as many as eighty thousand soldiers were passing through. His admiring parishioners at the far-flung churches he served began to chronicle his life as he lived it—they kept notes on his work, his whereabouts, his statements.

Then, on June 14, 1945, while he was in the small cowboy town of Lampasas, Roach opened that letter from Bishop Byrne:

"I am hereby appointing you pastor of the newly erected parish of St. Mary's (of the Miraculous Medal) . . . at Texas City. We have a map here in the Chancery from which you can get the outlines of your parish boundaries. There are one or two nice little cafés in Texas City, I had a meal in one of them not so very long ago. . . . You have a wonderful opportunity here, plenty of work, and I hope that you will be happy and successful in it."

THE DAUGHTER OF one of the most powerful men in Texas City, an oil executive who had brought the massive Pan American Oil refinery to town, decides as a matter of simple curiosity to see what the new priest is all about. She is there for Roach's first mass on the first Sunday in July 1945. As she slides into the wooden pew, there is a kicky optimism that hasn't existed for a very long time. The war in Europe has mercifully ended. The push to bring Japan to its knees is grinding forward. There is a welling sense that the world may be finally coming to its senses.

The oilman's daughter feels something that day in church, something that a handful of other people would also eventually admit to experiencing. The new priest, she very sheepishly says, has a glow about him. She has never thought much about the idea of someone having a sense of grace.

"Roach is a man of grace," she tells people.

In Texas City, she and others begin to talk about "the Roach glow"—something intangible about him, something that seems to suck you toward him. Sometimes, with one of her girlfriends, she parks near St. Mary's at a point where they think the priest can't see them. They peer over their dashboard, trying to grab a glimpse of him from a distance. He smiles when he walks out of his cottage or takes a seat on the steps of St. Mary's.

Roach talks with the oilman's daughter. She has been raised with religion but has moved away from it. After they talk some more, she tells her family that she is going to abandon her job as a secretary for a tugboat company:

I want to begin a new life as a nun. I want to be assigned to live in the most primitive, isolated religious community available.

Her father's firm, Pan American, is one of the biggest crude oil producers in the world. It is partnered with the American Oil Company (Amoco), and its roots go back to the ultimate, founding powerhouse of oil in the United States: John D. Rockefeller's Standard Oil. Amoco has a major stake in Texas City—and exists there without having to pay any taxes.

Now the daughter of the man who brought Amoco to town wants to join an isolated religious community. She wants to be as close as she can to the saintly Bill Roach—even if Bill Roach is demanding that her father's firm be annexed and taxed, even if he is railing about a "steel belt" choking the life from the piss-poor dockworkers, gang crews, and sailors packed together on the Texas City waterfront.

She isn't alone. Since Roach arrived, there are other enraptured young women who want to hear him speak. One of them, just like the oil executive's daughter, also wants to change her name to Sister William in honor of the new priest.

She's convinced that he is a saint come to heal the people of Texas City.

But now that his infamous letter has run in the *Texas City Sun*, now that he is demanding that the big companies get annexed and taxed, there are people who would really be happy if Bill Roach were dead.

Something real bad is going to happen to that fucking cocky son of a bitch.

The Ship

THE RUSTED Grandcamp *is finally escaping Houston, aiming for Texas City, and breaching that unnatural fog bank walling off the northern end of the bay.*

The heavy air, packed with the choking smell of chemicals, feels oily as the sailors desperately lean out and try to stare their way through the haze. The forty sailors are more than edgy and weary. Texas has been hard—that bitter chemical flavor is everywhere, rising up off the dismal water in clouds that make skin your blister and your eyes water.

When they had finally arrived in Houston four days earlier at City Wharf 4, Andre and his mates loaded peanuts, cotton, oil-well machinery, airplane parts, and seismographic equipment onto the Grandcamp. *Then Andre and his bone-tired friends were told that they would have to sail another forty miles back down the bay, to Texas City, to load the last bit of cargo—several thousand tons of ammonium nitrate.*

The sailors were grumbling: It didn't make sense for their Texas cargo to be split between ports that were so close.

Andre and the others had no way of knowing that the Port of Houston refused to handle ammonium nitrate.

Months earlier, private warehouse operators in Houston had seen endless bags of fertilizer passing from their railroad terminals onto cargo ships. Something about the shipments didn't seem right. They were never marked or flagged with any information. Two suspicious terminal operators hired a chemist to slice open a bag and analyze a sample. The chemist's report indicated that the compound was an extremely dangerous version of ammonium nitrate and that it had to be contained, at the very least, behind a firewall.

By 1947, the conflicted history of ammonium nitrate could no longer be ignored by government and military leaders across the world. They either knew that extraordinary history—or should have known the history:

At the dawn of the twentieth century, the most influential scientists in the world were convinced the world, as they knew it, was ending. Sir William Crookes, president of the British Association for the Advancement of Science, issued a desperate clarion call:

> ... All civilized nations stand in deadly peril of not having enough to eat. It is the chemist who must come to the rescue of the threatened communities. It is through the laboratory that starvation may ultimately be turned into plenty. ...
>
> The fixation of nitrogen is vital to the progress of civilized humanity and unless we can class it among the certainties to come, the great Caucasian race will cease to be foremost in the world, and will be squeezed out of existence by the races to whom wheaten bread is not the staff of life.

In Germany, the plea was heard by the brilliant chemist Fritz Haber. He quickly retreated to his laboratories and miraculously created a way of "fixing" nitrogen—of re-creating nature's own cycle of life and decay that provides the essential natural fertilizers and nutrients that are needed by almost every living thing. Haber, in the end, was out to perfect on heaven's own coda—he was out to re-create the great life cycle. And when news came that Haber had succeeded—that he had somehow gone into the laboratory and mastered that cycle with man-made fertilizers that could feed millions of people across the planet—he was seen as a savior. The announcement that Haber had won the Nobel Prize in 1918 underscored it all:

> Ladies and Gentlemen:
> The Royal Swedish Academy of Sciences has decided to confer the Nobel Prize in Chemistry for 1918 upon the Director

of the Kaiser Wilhelm Institute at Dahlem near Berlin . . . Fritz Haber.

Haber's (method) is of universal significance for the improvement of human nutrition and so of the greatest benefit to mankind.

We congratulate you on this triumph in the service of your country and the whole of humanity.

Millions of pounds of it began to be produced in chemical plants around the globe—especially when Haber and other scientists perfected ammonium nitrate's twin capabilities: as the most perfect life-sustaining compound, and as the most perfect, deadly explosive. Haber, who also oversaw the first massive "poison gas" releases in World War I, watched as the world became quickly addicted to his processes—despite an endless number of nightmares:

In September 1921, forty-five hundred tons of ammonium nitrate blew up in the German city of Oppau—the huge plant, run by Fritz Haber, was literally lifted off the ground; as many as one thousand were presumed dead; the city was virtually destroyed. It was the greatest man-made disaster in German history.

Three years later, forty-eight hundred pounds of ammonium nitrate exploded at Nixon, New Jersey. On April 29, 1942, in Tessenderloo, Belgium, 150 tons of ammonium nitrate exploded and killed a hundred people. In Milan, Tennessee, in February 1944, ammonium nitrate detonated at a government-run bomb-making plant and killed four men and injured another seventeen.

That latest explosion pointed to those millions of pounds of ammonium nitrate pouring out of bomb plants all over the United States during World War II. Five-hundred-pound airplane bombs, mixed roughly with two-thirds ammonium nitrate and one-third TNT, were being routinely dropped on Germany, Japan, and occupied nations. In Texas alone, at the enormous Pantex plant near Amarillo, bomb makers produced 1,019,200 pounds of ammonium nitrate.

World War II was fought with ammonium nitrate. And when the war ended, Gen. Douglas MacArthur issued a blistering ultimatum to Presi-

dent Harry Truman, essentially demanding millions more tons of ammo-nium nitrate—to win a different, new war. MacArthur, in charge of the forces occupying a conquered Japan, sent an urgent, top-secret memoran-dum directly to Truman and General Dwight D. Eisenhower in 1946. He wanted either "bread or bullets" to maintain order, squash terrorists, and thwart communism:

TOKYO, 1946
From: Supreme Commander Allied Powers
To: War Chief of Staff

OUTGOING MESSAGE—SECRET

"... THERE WILL BE WIDESPREAD STARVATION AND UNREST ... EITHER FOOD OR SOLDIERS MUST BE BROUGHT ... AND EITHER OR BOTH SHOULD BE STARTED IN THIS DIRECTION IMMEDIATELY...."

In Washington, President Truman and his advisers were analyzing end-less reports that would form the blueprint for the Cold War. In the wake of World War II, dozens of countries around the world were in play—the lin-gering battle that would pit the Soviet Union against the United States was gathering steam. Truman decided that the blustery MacArthur was right. Whole nations needed to be conquered again—this time with the food that would seduce them away from the Soviet sphere.

Following Truman's edicts, orders were issued through the Secretary of War to reopen fifteen U.S. Army ordnance plants for immediate production of ammonium nitrate. Thousands of tons were once again urgently cranked out in plants guarded and supervised by platoons of heavily armed army sol-diers. Much of it was destined for the docks in Texas—and for oceangoing vessels like the Grandcamp *that would carry it overseas.*

But, in late '46, in Houston, port supervisors reached a decision: Their port wouldn't handle ammonium nitrate.

Texas City, just down the bay, would.

The people who ran Texas City, the people who created it, had always offered up the city. Texas City had always been a willing lubricant for the government, the military, the oil companies, and the chemical companies. That was the reason it was founded. That was the reason it existed.

The Grandcamp *would dock there—and so would any other enormous freighter that needed to transport ammonium nitrate.*

This night, his ship surrounded by blinding, rancid fog, Pierre Andre is belowdecks on the Grandcamp *and thinking about his wife and daughter in France, the ones who had zealously guarded his tracks, his boot prints, in the mud outside their house.*

Maybe, if he is home in time, there will still be spring blossoms.

Now his battered-looking ship has entered the deep-water channels, approaching the Texas City waterfront.

The Mayor

APRIL 10, 1947

It was City Commissioner L. C. Dewalt who followed a black man down to where the shadows spilled out from the refineries—and then shot him in his fancy car.

It was back when the physically imposing Dewalt had served as chief of police. Curtis Trahan, who now sat next to him on the City Council, knew what Dewalt had done. Bill Roach, who was spending half his time in the black neighborhood, knew it. Most people in town knew about it, including Frank Anizan, who worked on the petroleum "cracking" line at Amoco:

"Down there in what we call Nigger Town. Down south of Texas Avenue. There was all kinds of red-light districts down in there. They had lots of sailors come in and they'd always hang out down in there. . . . Dewalt killed a nigger down there. . . .

"They had some trouble with him—he was considered a bad nigger. Dewalt told him to take it easy. He had a Buick with wood-spoke wheels. Dewalt told him not to drive through town like that. He come through it like he was going to a fire. So Dewalt got after him . . . pulled up behind him and got out and said: 'C'mon Davis, I'm going to take you in. I told you not to come through town like that no more.'

"And that nigger told him, he said, 'I'm not going anywhere with you,' and he reached in the back of the car. Well, he just shouldn't have done it

because Dewalt jerked that .38 automatic pistol out and shot him five times right over the heart.

"Dewalt didn't allow them across Texas Avenue, north of Texas Avenue at all. He'd catch one on this side of Texas Avenue, he'd say go back on your side over there."

Now, City Commissioner Dewalt is the closest ally at City Hall for The Company. For Mikeska. For the oil refineries, chemical plants, and defense contractors.

Dewalt is also Bill Roach's and Mayor Curtis Trahan's biggest enemy.

Dewalt is dead-set against Trahan's and Roach's plan for annexing and taxing Amoco, Union Carbide, and the other corporations that own and operate Texas City. Dewalt knows that Trahan and Roach will probably use the money to clean up The Bottom—but plenty of people feel it really isn't worth cleaning up.

It would probably be better if it was all just blown up.

CURTIS TRAHAN HAS been in office exactly one year and one day.

Now, sometimes with no specific mission, Curtis drives down past Clark's Department Store, past the Corner Drug Store, heading south toward the Negro quarter. In The Bottom, he stops by Eva Wilson's house, where he drops off his two little boys so Eva can baby-sit them for the day. Sometimes Trahan parks his new car on Second Avenue, near Rev. F. M. Johnson's First Baptist Church, and just walks around.

There is the endlessly thrumming echo from the refineries. The refineries and the chemical plants and the tin smelter are like abrupt volcanic mountains in the ocean; the homes in The Bottom are like insignificant, wobbly boats tied up at the feet of those mountains. Curtis puts a foot on the front steps of Norris's Café and listens to what people have to say. And he can see, down the block, Bill Roach walking with his head down and his arm around longshoreman Ceary Johnson.

Roach got over it a long time ago, that guilty feeling that he was acting like some kind of holy anthropologist coming to visit the poor natives. Roach didn't ever seem to have an ulterior motive. At least not one that Curtis thought he could detect. For his part, Curtis never thought much about the symbolism of the white mayor showing up in Nigger Town.

"My father never spoke an unkind word about a black person. Never. My father was uneducated, but he was intelligent," Curtis once said. "I really don't know how he had that kind of feeling but it got transferred to his sons."

Trahan is six feet four and has on his oblong face a slice of a smile that mysteriously suggests he is tapped into a nourishing, constant secret. Years ago, before he went to war, he had a tendency to erupt. He scared the living hell out of people who wondered exactly which Curtis Trahan they were going to get that day. Now people find it soothing being with him, even if they can't say exactly why. They like it when he is around, even if they have no definite reason for being in his company.

At Bob McGar's service station, they say it's because the new mayor of Texas City isn't fixated on reassurances. Others say the thirty-two-year-old Curtis radiates calm in an unsettled place like Texas City where so many people feel disconnected and dwarfed by the railroads, plants, and refineries. His wife, Edna, says it's only because he gives away anything to anybody who asks for it.

"Curtis is a steady man," says a Republican Oil refinery grunt who, like Curtis, had narrowly escaped coming home from the war in a box.

Really, that was all that needed to be said:

A steady man.

HE HAD STARTED this day, the most important one in the history of the city, as imperturbable as ever.

The annexation-taxation meeting was set for four-thirty P.M.

Gulping his coffee, he could look outside and see that the morning was dusted with another dull-turquoise chemical canvas just below the clouds. With the sun rising, the sky looked as if it were clotted with clumps of aging cotton candy. But the temperature was tolerable, and his wife, Edna, had propped open the windows in their home at 815 Eighteenth Avenue North. With Texas City averaging sixty inches of rain a year, she liked to take advantage of any clear day to open the windows all the way.

"Edna is a fresh-air fiend," Curtis always joked with John Hill, the young Union Carbide engineer who would bring his wife over to play cards.

From overhead, the homes in Curtis's neighborhood are like a child's set of blocks neatly arrayed row after row: The homes have two windows at each of the four corners; each house is set on concrete piers about eighteen inches off the ground—a nod to the fact that Texas City is almost at sea level. There is a semicircle-shaped stoop and a little overhang over the front door. In the back, there are four wooden steps leading to the yard.

Bustling through his house, Curtis searches for his hat and the keys to his car. He has a schedule:

Go to the city barn, the place where the city workers—the street crews, the dogcatcher, the mosquito-control man—were waiting for any instructions before they made their rounds. Go to City Hall to check messages. Maybe time for coffee with Roach?

In the last few months, he and Roach have been arriving at the same controversial conclusion from different directions: Although there is money, enormous, ungodly amounts of money, being made in Texas City, not enough of it is going back to the honest, hardworking laborers who built the place, who lived and slaved there.

For weeks, Curtis has been catching hell from The Company, Dewalt, and the on-the-ground officials of the national companies that have their billion-dollar plants in Texas City. Dewalt and especially Mikeska have been on the phone, in Trahan's little insurance office, threatening and demanding that he come to his senses and stop listening to Roach.

Curtis, you don't understand.

Curtis felt he was following some obvious instinct. He had even heard that other towns on the Gulf Coast had put together clandestine, midnight plans to annex their own industrial areas.

Hell, it seems so simple.

Everyone can see that there is a windfall in taxes to be had—Monsanto, Union Carbide, the Rockefellers, and all the Texas oilmen can easily afford it. Texas City has already given too much away. The land to build the tin smelter had been sold for a dollar. Monsanto had scooped up the old sugar refinery for just over $400,000 and turned it into one of the world's leading producers of plastics. The war had made each and every business on the waterfront enormously wealthy. Meanwhile, of course, Texas City had some of the poorest neighborhoods in America.

"I just stirred up a hornet's nest. Some of the businesspeople . . . some of them . . . unmercifully would chew on me. A lot of the businesspeople here just thought I . . . ," Curtis would say until his voice just trailed off.

He didn't have to finish the thought.

Dewalt and Mikeska framed it for him and everybody else in harsh terms: Curtis Trahan and Bill Roach are going to kill Texas City if they push their plan.

"The plants are gonna just pack up and leave."

Lately, Curtis has been thinking about how he has years invested in Texas City. There are big milestones to mark his time there. He met his wife in Texas City. He got his first job in Texas City. His two young boys were born there.

Curtis always assumed he would stay in Texas City forever. Probably until he died.

HIS FATHER WAS a Texas hard-ass, a muscled laborer who specialized in ramming together piers and docks up and down the ragtag coast—working as the "derrick man" on a pile-driving rig and stabbing thick wooden poles into the mad mess of intersecting bayous, rivers, and brackish shore waters. His parents divorced when he was five, his father took him in, and it seemed as if Curtis had been working from that day on.

After high school, Curtis signed up for an entry-level slot as a sample man at the Pan American refinery in Texas City. Every morning, he waited for the refinery bus to pick him up. At the plant, he was handed a tin bucket loaded with empty glass bottles. Curtis entered a chamber with impossibly high ladders surrounding huge cauldrons of oil ready for refining. He'd bound up the ladder with the other refinery workers, pausing to marvel at how he could take three steps at a time. At the top, Curtis lowered the bucket into the liquid and let the chemicals seep into the bottles. He delivered the noxious mess to the highly paid men in the laboratory.

When he was twenty-one, on the first day of May, Curtis found a reason to keep working at the hellholes in the refinery. Years later, when he was away at war, not sure if he would walk again, it would all come back to him:

Watching Edna driving with one of her girlfriends, slow and easy, down Sixth Street . . . his standing on the corner and watching as the girls suddenly waved and asked if he wanted a ride. "Where are you going?" Edna's friend wanted to know, and he had smiled and said: "Oh, I'm not really in a hurry to go anywhere," as he climbed in the car. Then it was as if Edna and Curtis were alone in the car. He could tell that she was worried about the way she looked, the way she kept moving the mirror so their eyes couldn't connect with each other, the way she smoothed the folds in her dress (she told him later that she had been embarrassed because it was just an old housedress) . . . and his listening as Edna said, somewhat mysteriously: "Do I bother you?" . . . He knew then that she was worth knowing . . . and for some reason he kept adjusting the windows so the Gulf Coast air wouldn't blow her hair where she didn't want it to.

A few days after that first car ride, he saw Edna standing in a driveway. Curtis had a funny question for her: *Do you object to dancing on Sundays?*

Edna didn't, of course. That night she borrowed her sister's white organdy dress designed with bold red dots and frilly butterfly sleeves. They went to a surfside place that featured live music, and they danced for hours. For weeks on end, Curtis just happened to be there when Edna walked out the door on her lunch breaks from Stead's Chevrolet dealership.

At the Armistice Day parade in Texas City one crisp November morning, Edna was the hometown beauty behind the wheel of the late-model Chevy that her boss had draped with the dealership's name. Every turn she made, Curtis was standing on the corner. He was sprinting from one to the next, waiting for Edna to pass, acting as if he had been at each intersection the whole time.

The other Texas City girls began to piece it together.

Edna finally admitted that they were going to get married a week before Christmas. By springtime she was pregnant and happily accustomed to his quiet ritual: every two weeks when Curtis got his oil refinery paycheck, he walked downtown, cashed it, and then made an even longer walk out to the edge of town and past the city limits. He'd hitchhike to Houston, where his father now lived, out of work and crippled. Every payday, Curtis would thumb a ride to give whatever money he could spare to the man who had raised him while still managing to do the bone-deadening kinds of work that had turned the untamed Texas coast into a place where the big ships could dock, where the oil could be transferred, where the chemicals could flow.

Edna wasn't surprised.

When she was about to go to the hospital, about to have a cesarean section to deliver their second son, Curtis had gotten a phone call. One of his refinery buddies desperately needed to borrow cash.

"Come on over and hurry it up," said Curtis.

AFTER EIGHT YEARS, one month, and twenty-two days at the oil refinery, he had had his fill of the nauseating aroma in those non-air-conditioned sheds where the heat made the fumes so intense you'd sometimes lose all feeling in your face. He had joined the union, the local branch of the CIO that covered the oil and chemical workers. He had gotten involved in the always dicey game of union politics in Texas and had gone on to serve as the secretary of the local. Curtis appreciated having work; he knew he was lucky to have a job, but he always felt that the companies needed to do right by their employees.

And now Curtis hadn't been to college, but he knew enough to know that he wanted out of the refineries. He also knew that people needed insurance. He decided to make a go of it as a local insurance man, at a time when the insurance man was considered your friend. Curtis saved money, got the licensing, and operated out of his dining room. Later, he found a small office in his stepfather's real estate agency. He cultivated friends and neighbors as customers—at the Episcopal church, at Lucus's Café, at McGar's garage. He found that if he did what he liked to do anyway—listen to people—he would almost always wind up with their business.

Early on, Curtis told Edna he had come to another decision: "I don't like going into people's homes, occupying space on their couches, and talking about the possibility of death," Curtis told her. It wasn't in his nature. He decided to concentrate on property insurance, not life insurance.

Besides, there were going to be storms. Curtis knew that after growing up on the coast. There was always uncertainty. Insurance seemed like the right game. Everyone needed it, especially in the summer of '43. The big war was on. The uncertainties seemed as if they would last forever.

That year and the year before, dozens of German U-boats were snaking underwater in the Gulf of Mexico. With few people even aware they were so close to the U.S. coast, the German submarines sank fifty-six merchant ships. Top-secret orders were issued to ratchet up security from New Orleans to Corpus Christi, especially in areas that had been turned over to the war effort and the race to make rubber, ammunition, oil, and fuel. Turrets for long-range guns were built, and army squads were dispatched to strategically important areas, including almost nine hundred soldiers who began patrolling the waterfront in Texas City.

Then, on July 27, 1943, meteorologists in Galveston, Texas City, and Houston knew that a major, serious hurricane was gathering strength over the warm Gulf waters—but because of the U-boats, the meteorologists were ordered to consider it a military secret. No storm warnings were broadcast to the people of Texas City. There was no call for evacuation or precaution. The government never warned its citizens.

The ferocious storm bore down on Texas City and trees of lightning, hundreds of them, stalled overhead. The official gauge in Texas City measured the winds at 104 miles an hour. The storm mass seemed never to stop rolling over Bolivar Peninsula, taking dead aim at Texas City. The rain whipped violently, horizontally. The old Southern Hotel, where the oil barons liked to stay when they took their private railroad cars to check on their Texas City operations, was demolished. Grade schools were condemned. As always, the first homes to crumble were the flimsy ones for the people exiled to live close to the waterfront. It was the worst hurricane since the biblical ones in 1900 and 1915.

For days, cradling cups of coffee at Lucus's diner, cynics said it was all too perfect.

No U.S. government warnings for Texas City.

The powers that be who owned and operated Texas City—the government, the military, and the corporations who had their headquarters in other parts of the United States—had forgotten that people lived there, too.

After the storm, Curtis was out on Sixth Street, striding to his car, fumbling with the paperwork sticking out of his briefcase, stopping to jot something down on the back of a binder. Curtis was going door to door, processing as many claims as quickly as he could, figuring that he had an obligation to everybody who had helped him escape the refinery.

Curtis had been right.

There would always be nightmarish, deadly storms in Texas City.

The Ship

APRIL 10, 1947

THERE ARE *floodlights playing across the docks.*

In the shadows, the longshoremen are huddled together, gulping thick black coffee, talking in low tones and staring expectantly out toward the choppy bay. The slick water is slapping against the oil-stained concrete pilings. Lines of thick string, left behind by local fishermen and baited with chunks of freshly killed chicken that might attract some crabs, rise and fall with the swells. When the wind howls, the light poles ringing the docks sway and the tin roof creaks at Warehouse O.

Finally, the Grandcamp *takes shape as it moves through the misty turning basin. You can see the outlines of sailors scrambling onboard. You can see the bobbing little orange glows from their cigarettes.*

The lumbering freighter is charting its final approach.

Onshore, veteran wharf master Clarence Green has given the all-clear signal to the ship. He wants it guided into Pier O. Dwarfed by the ex–Liberty ship, gnatlike tugboats chug off, aiming for the Grandcamp.

The wharf master watches the specter of the ship moving out of the fog and welling up as if it is being squeezed from an enormous mass of gray cotton. There is an almost utter silence as the freighter bobs ever so slightly and knifes its way closer to Texas City.

The Mayor

APRIL 10, 1947

BACK DURING the middle of the war, Curtis knew his draft number was coming up. He no longer had a deferment from working at the defense plant/oil refinery. He wanted to go to war. He was almost embarrassed that he was in his late twenties and everyone else seemed to be going overseas.

"I'm going to have to go," Curtis simply said one day to Edna.

He and a friend went to the recruiting station. He was twenty-eight, and he wasn't exactly color-blind. He preferred to think that he was missing a few shadings here and there, maybe a couple of nuances from the usual palette. But Curtis couldn't pass the damn navy eye exam. When he went to take the army eye exam, however, he was told that he had passed their test and that he was going to be inducted at palm-tree-lined Fort Sam Houston in San Antonio.

When Edna drove him down to the Texas City bus station, watching Curtis leave for God knows what, it was the worst day of her life. She had grown up in Texas City. Most of her family were living nearby. Like Curtis, she assumed she would stay there forever.

There was a lulling predictability: gathering at her sister Inez's house for the holidays, the kids eating at the card table set up in the kitchen while the adults gathered in the dining room. Everyone had their dishes—Edna would bring the big pot of chili, or the pumpkin, lemon meringue, or mincemeat pies. In the afternoon, the guys would cluster around the radio in the living room, listening to a Texas football game.

After Curtis's bus vanished from sight, Edna went straight to the Texas City dike. She drove out on the long flat ribbon, farther and farther out until it was as if she was alone on a landing strip set in the middle of the ocean.

Then she finally broke down and cried.

IN EARLY '44 he did infantry training at Fort McClellan in Anniston, Alabama. Curtis knew, without question, that he was headed for brutal action. His only brother, Ray, was already roaming with General Patton's Third Army. Before his final orders, Curtis was allowed to go home on furlough. At his small house on the north side of Texas City, on one of the final streets before the city limits, he walked across the threshold and his two boys scampered around his feet.

Curtis dumped the contents of his duffel bag on the wooden living room floor. The kids poked through it, disappointed that there was no rifle.

How could Dad be a real soldier?

Curtis spent a few days saying good-bye, and then he rejoined his outfit. He was a private assigned to the Second Infantry Division, B Company, 38th Regiment of the First Army. His shoulder patch had the 38th's motto, "The Rock of the Marne," in honor of the way the regiment refused to bow before a German onslaught in World War I. Curtis was shipped straight to the European theater and the Battle of the Bulge. He was "green," someone who was going to replace the dead in a devastated division, and like all "greens" Curtis was going to be watched closely: to see how he fought and, maybe more important, to see whether Curtis was an angel of good fortune or an agent of doom.

One Monday in January 1945, at 10:00 A.M., in the middle of the woods bordering a small Belgian lake, Curtis and four other men had somehow carved a hilltop foxhole out of the almost frozen ground. They rested for a minute on the dirt mound. From somewhere across the lake, moving hard and fast, a mad whirring sound came crashing through the woods.

The Germans had been unleashing hundreds of V-2 buzz bombs,

and there was no time to move, no time to react. This one, like so many others, had lost its guidance system and become a rogue explosive.

Atop the hill, Curtis was the only one hit . . . knocked off his feet and slipping into shock, he yelled for help and got no response. Bones shattered in his thigh, his left leg twisted backward, Curtis rolled onto his stomach, stabbed his rifle into the cold earth, and used it to inch himself down the hill. Blood gushed into the snow. Each time he lurched forward, his mangled leg took another twist. He waited at the base of the hill, fighting delirium, trying to picture Edna at home, everything about her.

She talked so much; she remembered everything in infinite detail.

Finally, in the frigid woods, the First Army medics arrived, their knives flashing as they cut his pant leg and pressed hard against the open wound to stop the bleeding. Curtis was shoveled onto a field stretcher. One of the captains from his division stared down at him.

"Boy, that's a million-dollar wound. That's your ticket home," muttered the officer.

Across the lake, a German patrol unit was watching. They studied the scene as two more American military vehicles hurtled toward where the buzz bomb had crashed. The German scouts radioed that the vehicles seemed to belong to high-ranking American officers. Orders were issued to lay down gunfire on the Americans.

Inside a metallic blizzard of bullets, Curtis's evacuation Jeep jolted through the woods, retreating to a medical tent. No one had thought to tie his legs down. They bounced high and came down hard with each pass over another rut. Curtis screamed in pain. He reached up and grabbed the soldier in the passenger seat. Curtis yelled for the damn Jeep to stop and for his legs to be secured to something.

In the field tent, Curtis watched grim medics talk in whispers about the best way to save his mangled left leg. He was sent to an army hospital in Paris. Outside his window were one tree and a cheerless, snow-covered courtyard. They would constitute his only memories of the city.

At home, Edna got a call from a girlfriend who worked at Western Union. There was a telegram from the War Department: *Private Curtis Trahan has been injured in combat in service to his country.*

Her brother Harry came to see her and took her to her sister Inez's house. While everyone stared, she sat in a chair and cried for hours. Finally, a doctor was summoned. He came to the house and gave her a shot. Still crying, she felt herself drifting into sleep and thinking:

We had the world on a string; we had the world on a string.

CURTIS WAS SHIPPED across the English Channel to imposing Woburn Abbey, the Duke of Bedford's gilded estate that had been turned over to a variety of military missions—including being the headquarters for covert propaganda broadcasts into Germany. From England he took his first and only cruise on an ocean liner: Curtis was loaded, with hundreds of wounded men, onto the *Queen Mary* and shipped back to New York.

Onboard, with the bandaged men in beds stacked three layers high, Curtis listened to the crackling broadcast announcing that President Franklin Delano Roosevelt had died. His eyes began to well with tears. And when Curtis dared to look around, every man, every injured soldier finally coming back home, was also crying.

For years, when Curtis allowed himself to think about the war, what formed crystal clear in his mind was the image of all those grown men and their poorly contained grief taking shape somewhere in the middle of the Atlantic Ocean.

Curtis spent five days in a hospital in New Jersey and then was shipped out to Fort Hood, near the small town of Temple, Texas. He underwent more surgeries and recuperated in another hospital for six months. The doctors had him in casts stretching from his foot to his hip. He was going to be laid up for seven months. Immobile, Curtis passed time by watching the soldier in the bed across the room build little wooden models of LSTs—tank landing ships. When the doctors finally told him he could go home, he called Edna.

"Let me come and get you," Edna said.

"Don't worry," said Curtis. "I'm going to hitchhike back. It'll be okay. I'll catch a ride to Houston. Meet me downtown. Meet me in the men's department at the Sakowitz store."

Edna drove up the diesel-stained bay road to Houston, aiming for the growing skyline. She found a parking spot on the wide Main Street and went into the big Sakowitz store. Edna walked to the men's department and waited. Hours passed. Finally a salesman told her the store was closing and she had to go home.

"My husband's coming home from the war," said Edna. "He wants to have a good suit before he gets to his house in Texas City. That's why he told me to wait here for him."

The salesman disappeared and came back with old man Sakowitz, the founder of the chain. He stared at Edna, turned to the salesman, and said: "You're staying in the store until her husband gets here. And you're going to give him anything he wants."

Finally, hours after Edna had arrived, Curtis appeared at the front door of the closed store. He was on crutches and somehow toting a duffel bag with a Purple Heart inside. The salesman let him in. Edna guided Curtis to a rack of suits. There was a nice navy one, made out of wool, that would have to be altered a bit. Curtis walked out of the store, into the humid night, with a damn fine-looking new suit that he would be wearing when he woke up the children at home. For weeks, Curtis wore his Houston suit, and he learned to lean on a cane for a while.

Back home, Texas City seemed electric, bustling, even more than he remembered. It seemed filled with possibilities. And when he was making the rounds to his old insurance customers, they noticed something about him. It was the same thing that they had seen in other men who had come back home after surviving something immensely cruel, elemental, and furious. His sense of ease was now almost wholesale. His volcanic tendencies were long gone.

Curtis Trahan was a steady man. He had the glow of a good man who really had denied death its dominion.

RUNNING FOR MAYOR was, like many recent things in his life, a turn of the wheel that was going to happen even if he had never dreamed it or desired it.

The notion had fallen in his path after some young, optimistic guys from the Lions Club, from church groups, and from the local longshoremen unions decided that there was only one person in Texas City no one felt threatened by: Curtis Trahan.

Curtis had hardly ever talked about his upbringing, his background, where he had come from, where he had been when he became what some people in town called a "bona fide war hero." His whole demeanor made people in Texas City want to compare him to someone they had seen in the movies at the segregated—Negroes upstairs, whites downstairs—movie theaters on Sixth Street.

They said that he looked like the young actor Jimmy Stewart. And when people began their comparisons, what Curtis usually did was lock his gaze on the floor, shake his head, push his hat back on his head, smile a little more. And if he was sitting, he would absently reach for his leg, the one that was almost blown off in World War II, and gently place his hand where the wound had left a ragged gouge.

Of course, it all reinforced the comparison.

But then, when he was out of sight, there were other people saying elaborate, conjectured things.

Curtis Trahan was getting cocky.

If you didn't know any better, the whispers went, you'd think that Curtis Trahan, running for mayor of Texas City, was actually smug instead of being very, very certain of some things.

He's sucking up to the unions and the longshoremen. He's talking too damn much to Roach, the Catholic priest.

And some people said that if you watched where he was going during the day, there was going to be a big, big problem.

He thinks there should be niggers on the police force.

At their home, Edna studied the way he did simple things, like moving about in the kitchen. Built like a football player and walking on a leg that had almost been blown off by a Nazi rocket, he liked cutting up apples in little slivers and tenderly laying them on top of his home-made bread pudding so that the apple slices would brown up nicely.

Edna said he was a human sponge, in the very best sense.

"My Curtis never met a stranger," Edna liked to say.

In early 1946, with the endorsement of Roach and union workers, Curtis was elected mayor in a landslide.

HE WAS GOING to get $50 a year. He was going to get an office in the old building on Sixth Street—he always thought it looked more like a community center, with an auditorium, a meeting hall upstairs, and the library (and the jail) all attached. He was expected to just cut ribbons when a new mom-and-pop store opened on Texas Avenue or Sixth Street. He was just going to suffer the town's eccentrics—and there were many—when they stood up at the City Council meetings and proclaimed their distaste for unleashed dogs. He was just supposed to ask moon-faced Fire Chief Henry Baumgartner to recruit more men for the volunteer fire department. He was going to let Police Chief Willie Ladish do whatever the hell he needed to. He was going to file the simple paperwork to get the county to pave some roads.

At Lucus's Café, where the Lions Club met, people say it never matters who is mayor.

Texas City is a malleable thing, a tool more than a community. It is a company town. It is a government town.

It's beyond the control of Curtis Trahan or any of his friends, including that priest Bill Roach.

THE HISTORY-MAKING ANNEXATION battle begins at 4:30 P.M.

Curtis has already been making people uneasy over the last few months. He nominated the first black man to sit on a grand jury in Galveston. He had gone ahead and named Willie Cole the first black police officer in the history of Texas City. And he heard the taunts: *He had all the nigger women running through town with their dresses up and shouting: "Vote for Curtis Trahan!"*

City commissioner and ex–chief of police L. C. Dewalt, the man people had learned to fear down in The Bottom, is ready to tackle the issue. He wants to talk first.

This is the greatest issue to ever face the city. This session should be immediately adjourned so we can study the issue at a later date.

Dewalt looks around the room.

The city commission had always been against annexation, and that's the way it should stay. "It's taxation without representation."

Curtis promptly votes Dewalt down. Two other commissioners follow his lead. Looking on, Bill Roach is almost dizzy. He stands up, asks to speak, and is recognized by Curtis. Roach begins slowly:

"A healthy social environment will make men happy and they will produce more." Texas City isn't going to bleed the corporations. Ceary Johnson, the longshoremen down at Frank's Café, Trinidad Garcia—they all need to live in "frugal comfort."

Mike Mikeska can't stand it. He's been counting on Dewalt, on someone to destroy this whole goddamn business about annexation and taxation. Mikeska, president of The Company, stands up and takes his turn:

I'm just a private citizen, somebody who has lived here for thirty years, unlike some newcomers.

Curtis and everyone else has to know that the "newcomer" remark is aimed at Bill Roach. Then, Mikeska makes sure to mention the names of most of the important companies that have a stake on the waterfront. The oil companies, the chemical companies are the backbone of the entire town. They provide every single damn job in one way or another.

Look at Amoco, it was a "savior" when it got here.

Mikeska fixes his gaze on Trahan and the commissioners.

I want you to delay this vote.

The comments of the most powerful man in Texas City are duly noted. The meeting rolls on. After an hour of speeches, all eyes finally turn to Mayor Curtis Trahan. He really does seem like Jimmy Stewart as he talks in his earnest, uncomplicated voice:

"Industry should have no more objection to taxes than the natural objection we all have to paying any tax. If we govern fairly and tax fairly we will not be in any trouble. Industry will not be stifled and the people will benefit greatly."

When he is finished, Curtis says he has made up his mind. He is casting the deciding vote in favor of annexing the entire steel band, including Union Carbide and Amoco, with all of its connections to Wall Street, the White House, and beyond.

But, there is one more stunning announcement. There is something else, another motion, that at first almost seems insignificant compared to the pivotal annexation-taxation debate. Curtis wants the commission to authorize him to do something else:

"To begin immediate work on a successful solution to the crowded situation in the Negro and Latin American areas of the city."

Roach watches in amazement as the motion authorizing Curtis to make life better for blacks and Mexicans also passes. And then, just like that, the meeting is being adjourned and city attorney Bill Dazey is scrambling to offer some final words:

You are split in your opinion, but you voted your conscience, you voted what you thought was right.

It is 6:00 P.M.

Curtis and Roach walk down the steps of City Hall knowing that the recriminations are going to fly to Monsanto headquarters in St. Louis, to every corporate office in Houston, to Union Carbide and to only God knows where else.

Curtis and Roach have succeeded in annexing the steel band—and now they are going to tax it. Now they are going to fix The Bottom and El Barrio.

Night is coming on, and the clouds continue to manuever high over Galveston Bay. Warm winds are gathering in the south, and there are random spikes of lightning over the sea. When each burst of light flashes down, there is a sudden burned bronze glow bouncing off the metal pipes and towers at the refineries.

Roach, wracked by his sleepless nights and the creeping visions, knows that it is extraordinary. In a way, what just happened in City Hall is the culmination of everything he had been fighting for. He is ready, in his own way, for whatever comes next. That bloody cleansing is almost at hand.

Curtis heads to his car. If rain's coming, maybe he'll beat it. Edna will have dinner waiting. The kids will still be up for another hour or two.

To Curtis, the gulf sky sometimes seemed impossibly infinite. Sailors who didn't know better risked being enraptured by what was overhead instead of focusing on the tricky spoil banks and shoals. There are, of course, hardened sea captains who made a regular living by ferrying out to the freighters, the tankers, and the cargo ships and guiding them into port.

He turns the key in his new car, a Kaiser, a model made by the same people who were producing Kaiser Aluminum. In his rearview mirror there are men in suits arguing in front of City Hall. Behind them, in the distance, the tank farms and the chemical plants look as if they are also huddling in the dusk.

Sometimes, Curtis wonders what it would be like to be one of those seagoing masters of the Gulf of Mexico.

Someone like barrel-chested Henry Dalehite, almost everyone's idea of what a ruddy Texas sea captain was supposed to look and act like. Dalehite, stout and cocky, was the recognized expert on this part of the Texas coast, someone who understood the currents, the winds, the banks, and the oyster reefs off Texas City. Dalehite and his wife, Elizabeth, lived through every howling hurricane in the area in the last fifty years—including the Great Storm of 1900 that buried most of Galveston Island under dark water and claimed eight thousand lives.

Curtis Trahan never went to sea like the Dalehites.

He also never thought he'd become the mayor of Texas City and enter into political alliances with Yankee priests. Truth be told, it had never really been his plan. The job, like some life-changing things do, just chose him.

He knew he had lived through the most intense storms of his life.

The hurricane of 1943.

The Battle of the Bulge.

The City Hall fight tonight.

There can't possibly be anything bigger.

AT HOME, HIS two boys, eight-year-old Curtis Jr. and six-year-old Bill, are getting ready for bed. He kisses them and holds on to Edna. She can probably see the City Hall drama on his face. Outside, the four spindly live oaks and the one tallow tree he and his father had planted are bending with the night gusts. Curtis has traveled enough to know that Texas City isn't Paris, France. But it is home; it is where he has decided to make a stand. It is a place where maybe things can be made whole and good.

Texas City, in its way, never seemed more beautiful—and more poised. The nation had developed an unwavering faith in the fuel, the plastics, and the exotic alchemy practiced in the mighty plants ringing the Gulf Coast.

Besides, there is that new war that Texas City and its workers would win.

It is the dawn of the Cold War.

And now the president of the United States and U.S. Army generals are scheduling Texas City to receive 150 million pounds of ammonium nitrate. The men and women in Texas City will ship it out on freighters destined for starving families in France, Italy, and Germany.

But for now at night, away from the docks, it is as if the town has turned into a carpet that is rolled up tight for the evening. Nothing moves out on the streets as the lights in each small house are extinguished.

At Bill Roach's bungalow, he is alone with his nightmares.

They are more real, more dizzying, than ever before.

The Warning

APRIL 14, 1947

New York City

LONGSHOREMEN IN New York are flipping the pages of the newspapers and reading about Milton Reynolds—who almost always is called "the millionaire manufacturer of fountain pens"—and his expensive attempt to beat the round-the-world flight record held by billionaire Howard Hughes. And, at the front of the meeting room in Manhattan, D. J. "Jim" Gavin, a brawny field agent for the National Maritime Union (NMU), is reviewing his notes, anxious for the usual nighttime membership gathering to get started. He has frightening news to deliver.

He has just returned from a fact-finding mission along the Gulf Coast. Gavin, a no-nonsense, square-faced man with a wide forehead, has been a labor activist for twenty-nine years. He spent a decade at sea and then was asked to be a maritime union officer, showing up when the NMU needed a presence to protect its members and to monitor anti-union activities on the waterfront.

Texas is always a tough case, the kind of place where Texas Rangers like to race their sedans straight at rows of striking longshoremen in Corpus Christi and slam the brakes short of plowing into the crowd. The driver will then angle the car a bit and suddenly a line of long-barreled weapons will be pointing out the windows... and a voice from inside one of the sedans will beg the strikers to make a move. In Corpus, south of Texas City, a Texas Ranger is infamous for pistol-whipping dockworkers while keeping his thumb on the hammer of his gun.

Longshoremen and merchant seamen learned not to call the police for protection. That usually meant that the KKK was going to be tipped off, and the KKK was wide open down along the southeast part of Texas. Some Klansmen assumed that any unions in Texas were probably run by Jews from New York City. During waterfront strikes in the late 1930s, at least twenty-five other union men were killed.

There are other problems: The FBI has been investigating labor leaders to see if they are furnishing Soviet agents with information that might help them blow U.S. refineries and chemical plants to kingdom come.

When Gavin arrived in Texas, he traveled to the ship channel stretching from Galveston up to Texas City and then on into Houston. He interviewed sailors and longshoremen. He studied the pace and flow of traffic along Galveston Bay.

Gavin has seen things as a labor official during a time in America when union men faced down police, the FBI, the KKK, and strikebreakers with baseball bats. He isn't prone to being an undue alarmist, unless there is something truly serious to contend with.

Monday, at the nighttime membership meeting in the National Maritime Union headquarters in downtown Manhattan, Gavin stands up and begins to give his official report.

"Ship owners are shamefully neglecting the safety of crews on explosive-laden ships . . . when two tankers pass each other in the channel, they are so close together that you can practically jump from one to the other. Should they scrape together, they can easily blow up."

The union members are shifting in their seats, paying more attention when he starts talking about Texas City.

It is a "natural" for a major disaster . . . maybe a gigantic explosion.

The Ship

THE GODDAMN rains had thundered down, and everything was behind schedule. The humidity was building and building until the air was like some sort of moist jackboot pressing from the sky and you could wring a spray of sweat straight from your work shirt.

At the exact same time that Jim Gavin is delivering his cautionary report, longshoremen Ceary Johnson, Julio Luna, and other members of their work gangs are tossing down their cigarettes and clambering aboard the Grandcamp. They were supposed to have the damn ship loaded already. But those rains had come, stacked permanently for a day or two, as if Texas City needed to surrender to the elemental, natural things that had once always defined it—the ruptures in the sky, the storms, the lightning.

Now, there was a break in the weather. Hugging Pier O, the ship seems pregnant, resting heavy in the water and already fat with the cargo that it had picked up around the world.

In Havana it loaded car parts. In Matanzas, Cuba, it took on fifty-nine thousand bales of sisal twine. Then, in Houston, it loaded 9,334 bags of shelled peanuts, 380 bales of cotton, and oil-well equipment. In Venezuela, it was supposed to unload something—sixteen cases of small-arms ammunition that the ship had first taken onboard in Belgium. As sometimes happens along unregulated waterfronts, the cargo of ammunition never left the ship.

Now Luna, Johnson, and the other Texas City longshoremen are divvying up the assignments to the anxious crews. They will have to race to hand-load those 5 million pounds of ammonium nitrate. The Grandcamp is scheduled to raise anchor on Wednesday night and sail for Brest, France.

The Texas City longshoremen have handled other slippery, hot, cracked bags of ammonium nitrate for more than a year straight. Since Texas City accepted its first load in January 1946, at least seventy-five thousand tons have passed from the trains to the 880-foot-long steel-sided, iron-roofed, wooden-floored Texas City Terminal Railway warehouses.

From there the bags are transferred to the docks and then to the ships. In order to avoid cracking them any more than they already are, no hooks are used. The bags are hoisted onto pallets, then onto the decks, and finally into the holds. Then the dockworkers strip off their shirts, lower themselves into the dank holds, and begin the bare-knuckled process of building stacks seven bags high.

The bills of lading indicate that the 51,502 bags in Texas City have been produced in three midwestern military plants: the Cornhusker Ordnance Plant in Grand Island, Nebraska; the Nebraska Ordnance Plant in Wahoo, Nebraska; and the Iowa Ordnance Plant south of Middletown, Iowa. Tacked to the inside wall of each of the boxcars are shipping papers signed by U.S. Army officers. Railroad shippers are sometimes required to affix red warning placards to boxcars filled with potentially dangerous cargo. There have been no warning tags on the boxcars that began roaring into Texas City over the last three weeks.

As it spills out of the punctured bags, some of the dockworkers scoop up the ammonium nitrate and hold it in their palms. The ammonium nitrate is shaped into tiny pellets. It is coated with Carbowax to ward off moisture damage. Tinged brown from minute amounts of clay added to prevent caking, it is packed inside six-ply moisture-proof paper bags. Each of the 51,502 bags in Texas City is marked the same way:

FERTILIZER
(Ammonium Nitrate)
32.5% Nitrogen
100 lbs. Net
101.5 lbs. Gross
1.6 cu. Ft.
Made in U.S.A.

Ceary Johnson, Julio Luna, and the other men put their backs into the work.

With the winds coming from the northwest at fifteen and twenty mph, it feels colder than the official fifty-five degrees. But the bags are unnaturally warm. Some of the longshoremen press their cold hands down onto the heated stacks. Some of them even press their bodies on top of the bags, feeling the warmth soak into them.

Some of the thousands of hundred-pound bags are scorching.

They are too hot to touch, as if they are about to explode spontaneously into giant rectangles of fire.

The Priest

It had turned brisk overnight, and the gauge outside Roach's cottage registered fifty-three degrees. He sometimes cracks the windows open a bit to catch a breeze, and the westerly wind would creep into his room. His thick black hair is tousled against the pillowcase. Roach, thirty-nine, has a well-worn medal of the Virgin Mary around his neck. He lies in bed for a long while, watching the sunrise form, feeling the humidity seep through the walls and then, strangely, recede. Outside, anyone in the city can see a blue and orange glow in the sky, a light like the evanescent kind just before a match dies.

Roach finally slides his feet to the floor. Usually, he loves going outside for a good toasty smoke in the morning. It's good to stand barefoot on a sidewalk that's deliciously cooled. It's a good time for ordering his thoughts, arranging the battle plan for the day, staring through the webs of smoke curling up from his cigarette, and watching the strengthening sun burn off the coastal mist.

Roach would no doubt step out the front door and into the gauzy dawn. There is often the sudden flap of a great blue heron arcing out of a small pond rimmed by salt cedars near the Texas City dike. Truck tires are crunching over roads paved with weather-beaten oyster shells that are the dull hue of bleached bones. Bits of radio news are drifting from open windows in the nearby houses—somebody named Jackie Robinson became the first Negro to play in the major leagues yesterday.

Roach wraps his arms against his chest, lights another cigarette, and studies the sky. To the south, fifteen blocks away, he first notices a witch's cape of paisley smoke beginning to rise up from the industrial zone on the waterfront.

By 8:00 A.M., the winds are cascading from the north, blowing down from Houston, bending the saw grass and the cattails along Galveston Bay. The humidity, close to one hundred percent a few hours earlier, has dropped down to the eighties. Sometimes the humidity remains so fixed at sunrise, so poised at the highest level possible before it rains, that old-timers look out their windows and laugh about how "it's raining from the ground up."

This morning there is a sweet, almost aching, reprieve.

The temperature is nudging into the sixties. It is, really, the only pleasant time of year in Texas City. People are eager to go outdoors. There is a zone, from April to early June, when that honeysuckle and the jasmine and the oleander are going to prosper.

On days that dawn like this, Roach can almost forgive Texas City's callused faults.

The breezes occasionally carry tastes of the open sea. The sky, except for the gathering plumes on the waterfront, is unusually clear. In the mornings, in the yards of the older homes, some perfume from the oleanders and the honeysuckle lingers as if the fragrance is embedded in the carpet of dew pearls. Gulls are working against the gulf winds, hovering like kites over the dike and watching a middle-aged construction worker named Teodoro Garcia bait his fishing line before casting into the pewter-colored water.

Roach wishes his brother Johnny was with him.

Roach wishes Johnny could tell him what to do.

When he walks through Texas City, sometimes it seems as if all he can see are the children.

The city has no money and the schools are so overcrowded that they are split into double shifts. Some kids begin their school day at noon; some kids finish that school day by noon. When they are not in class, boys and girls ride bicycles and homemade scooters to the docks to watch the volunteer firemen hose down burning bales of cotton or overheated engines. There have always been little, easily managed fires near the docks, small blazes with their subsequent towers of smoke.

Sometimes, even, lightning would snake out of the sky and lash into an oil storage tank . . . tongues of fire would dance upward, and a crowd would quickly gather as if it was a fireworks show. There would be cheering and applause as the volunteers tamed the blaze, and then everyone would go home, knowing they had seen something better than any Showboat Theater movie, better than one of the pumping jump-blues bands that would barrel into town from nearby Galveston and Houston.

Bill Roach would stand on one end of the waterfront and stare at the children cheering for the battling firemen. There they were, lumped into bunches on the mushy lanes lining the busy docks.

Sitting with their feet dangling over the barnacled pilings, juxtaposed against the hulking oceangoing freighters, they have the satiated, satisfied glow of innocent children peering at the night sky on the Fourth of July.

ELIZABETH DALEHITE, THE smiling wife of veteran sea captain Henry Dalehite, can see the children as she steers her car toward the Seatrain Terminal along the docks; her husband, one of the famous sailors along the Texas coast, is on the passenger side of the car and catching a few more minutes of sleep before they get to the waterfront. Up ahead she can see also see the strange smoke rising.

On Seventh Street, right behind City Hall, twelve-year-old Harold Baumgartner has just watched his father, the irrepressible and well-liked Fire Chief Baumgartner, leave their home. Bill Roach likes Baumgartner. So does Curtis Trahan. Most people like Baumgartner, except for the ones who hate the idea that Baumgartner has allowed some "Spanish," some Mexican Americans, to join his volunteer fire department. Like he does every morning, the smiling Baumgartner is waving to the two little girls who live next door, the ones who have made a morning ritual of running to their window to say good-bye to the fire chief.

Mike Mikeska, the domineering president of Texas City Terminal, has backed out of his garage on Ninth Avenue North. He's got to go to the waterfront, to the docks, to meet with Walter Sandberg—his aide, his understudy, the man he is grooming to take over The Company. The smell of the coffee his daughter Beth is brewing is still on Mikeska's mind.

On the north side of town, a twenty-eight-year-old chemical engineer named John Hill can barely walk. Hill had been a hot-shit home-run hitter in college and loves to remind people he'd deliberately skipped a chance to play for the New York Yankees. Hill's got the flu or some damned bug, and he's wondering if he should call in sick to Union Carbide.

Forrest Walker Jr. is finishing up his gym class, trying to figure one more time how the hell he had gotten roped into being the lead in the senior class play at Central High—and sometimes trying to figure out his relationship with his father, the boilermaker who had gotten a job at the Monsanto plant and dragged the family to Texas City. As he walks back to his regular classes, he can see the elementary school playground where hundreds of Texas City children gather.

Longshoreman and WWII veteran Julio Luna is bouncing on the seat of a car that's barreling toward Texas City. He has survived a Japanese torpedo and now all he's hoping for is that there's some more freelance work waiting for him, maybe hauling more of those fertilizer sacks onto the big French freighter that docked in Texas City last Friday.

Kathryne Stewart has said good-bye to her husband, Basil. Her husband kisses their two little kids and is going to his public relations office at Republic Oil. He said he's going to be a little late coming home today because he has something important to do with Mayor Curtis Trahan.

Rev. F. M. Johnson, who lives in The Bottom and down the block from longshoreman Ceary Johnson, is in the shower, thinking about the fact that he has to get up to Houston for some church business. People seem to want everything from him—advice, jobs, you name it. The problem, he's figured out after being in town only a year, is that there isn't that much to give in The Bottom.

Edna Trahan sees her husband, Curtis, loading their son Bill into the car and pulling out of the driveway toward Danforth Elementary

School. When she thought her husband was lost in the Battle of the Bulge, she used to fall asleep crying, and sometimes there were those words in her head that wouldn't go away: "We had the world on a string; we had the world on a string." Somehow her husband had escaped, bleeding badly and barely alive, and made it back home as a bona fide war hero.

Now, on calm, seductive days like this, she feels the way she did before the war.

Things are right again.

They have the world on a string.

In Pier O at the north slip, the crusted French freighter *Grandcamp* is sunk low in waters coated by swirling, raspberry-hued slicks. It's one of three oceangoing ships still in port. Jagged hunks of oily wood, sheared from a shrimp boat that went down off the Snake Island shoal, are bobbing in the bay and slapping against the hulls of the looming cargo ships. Foamy clumps of cigarette butts, fishing line, and cans ride the small swells.

At 8:12 a.m., small rolls of kelly green and orange smoke are shooting up from the ship and piercing the aqua palette of the sky. They are unlike anything anyone has seen. The neon rolls of smoke linger for a few seconds, begin to flatten into carpets, and then simply disintegrate. A few of the rolls curl inland instead of blowing out to sea. People lean out their car windows, stop on street corners, and look up from their front porches. It's like the afterglow from a spectacular fireworks show, the shimmering way the sky will light up with splendid, dancing colors.

Roach knows by now that there must be a fire somewhere on the docks. It happens a lot. A sailor flips a cigarette into a haul of cotton, or a welding spark sets off some dried piles of lumber. Refinery crews and local firemen always immediately smother the blazes with foam and water.

By 8:15 a.m., the nectarine columns of smoke are surging over Frank Scofield's café, the waterfront dump squatting twenty-five yards from the pilings at Pier O. The smoke caps the sky above Frank's, and fingers of it point one block away.

Florencio Jasso, a World War II veteran who ships out on merchant marine ships, is walking through El Barrio and going up to the bus stop, away from the gathering smoke. The neighborhood is humming: Old Man Rizzo is arranging grocery trays filled with fresh-made tamales. Seven-year-old Freddy Vasquez watches his grandfather walk to the family's El Charro Restaurant. Trinidad Garcia, who once rode alongside Pancho Villa but is now a janitor and a lawn cutter, is watching the smoke and lifting the petals on his favorite roses as if his finger is crooked under the chin of an infant.

Now the thick cushions of smoke over the waterfront are unlike anything Roach has ever seen. The winds are now coming from the northwest at twenty mph. There are lines of laughing, running children headed toward the docks. The flapping clumps of smoke hover above the water, absorbing the sparkling vapors rising to meet it.

Roach stubs out his last cigarette and tosses it to the ground. Whatever has been holding on to him—intuition or visions—is now indelible and commanding, as if it is moving through his cells. Parishioners remember him saying out loud:

"This is it."

He steps back inside his cottage and slips on his clerical collar and black robes. Roach dashes to his car at the curb and jumps into the old Ford. No time to walk. His tires kick up white dust as he speeds east toward Third Street and turns left onto the narrow road that will take him to the docks. As Roach drives south, he would no doubt see a small parade of people, including the children, moving in the same direction. Mothers with baby carriages are pointing to the smoke and pushing through the crowds to get closer. A policeman has pulled his motorcycle to the curb to stare upward. Cars are darting down the unpaved back alley that delivers you right to the docks.

Sometimes when he drives, Roach will have his sleeves rolled up, maybe a cigarette lighter in his hand, and he will stick his head out the window to let the hot wind blow across his face. He is on one of the wide alleys to the waterfront. Ahead of him, he can see men shoving open the screen door to Frank's and hollering for the people inside to come out and take a look at the huge balls of smoke. He can see gangs of men moving toward the *Grandcamp*.

Roach caroms over the ruts. He angles his car alongside the diner. Thick squads of gulls and pipers are lifting off and moving away from the dock, frantically escaping out over the water and toward the oyster reef, the sandbar, Bolivar Peninsula, and the endless expanse of the Gulf of Mexico.

Roach could see the faces of the people going into Frank's. At best, Frank's is a waterfront way station for gossip, acidic black coffee, scrambled egg sandwiches, and stale doughnuts. Already inside are a dozen longshoremen, French-speaking sailors loading up on cigarettes, and a knot of bleary-eyed alcoholics waiting for "grab-out"—waiting for someone to grab them for a ditch-digging or concrete-pouring job that has to be done that day.

Roach frantically looks for Mayor Curtis Trahan, for someone in charge, someone from the Coast Guard, maybe Mike Mikeska, president of Texas City Terminal, that iron-vise private firm that for five decades has controlled the movement of the trains and ships in and out of the waterfront.

There's no one in Frank's whom he needs. He steps back, pushes hard past the men bunched outside the diner, and sprints onto the walk that takes him the twenty-five yards to Pier O.

It's 8:45 A.M.

As he runs, he passes longshoreman Ceary Johnson briskly walking in the other direction. Johnson stares at Roach and thinks that he's a good man, about the only damned white man you could seriously trust in Texas City. In a way that he hadn't ever talked with anyone about, Johnson had come to love Roach.

The priest is past Johnson now, scrambling onto the wharf, feeling the walls of heat and aiming for a well-dressed man who is staring into the ceaseless banks of smoke heaving from deep inside the superheated freighter. There are men screaming something from the gunmetal gray deck railings of the *Grandcamp,* their words being swallowed by the roaring, painful, cracking noises ricocheting deep inside the ship's holds. Roach watches as the ship's deck suddenly swells. It rises like a pregnant mass covered by hissing wood and hot steel. The water in the bay is beginning to churn, frothing as if it is being beaten.

Roach bolts to Mike Mikeska's side, and for a second the most important person in Texas City, the man who runs The Company and the waterfront, pulls his attention from the spreading commotion, the cresting noise, and the urgent voices on the pier. Dozens of firemen are racing into the smoke. Six firemen are futilely wrestling with a hose. It looks as if they are spraying steam instead of water—the inferno blasts are so intense that as the water roars out of the hose it is instantly vaporized. Mikeska turns and stares at the wild-eyed priest. They have been at war over the future of Texas City. Mikeska wants it the way it has always been. Roach wants it to change. Mikeska wants the big companies protected, buffered. Roach wants them held accountable.

Now tarpaulins of smoke are speeding and flooding across the fifty-six-foot-wide deck of the big ship. Pooling in corners, they spin into thick masses and soar straight into the sky. There is a nauseating chemical stench. Sailors are yelling in French and disappearing amid the billows. Frantic longshoremen suddenly come shoving out of a north-facing wall of black haze. Now there are twenty-eight firemen facing the ship, all of them desperately trying to unroll the hoses. They inch toward the huge, rattling freighter while dozens of sailors and dockworkers scramble down ropes and gangplanks and flee in the opposite direction.

On the rim of the docks, the wide-eyed children in the crowd of five hundred onlookers are hurrahing and clapping with each new commotion, with each new cloud that rockets up, spreads out over the bay like a painting by Matisse, and then zooms to the Gulf of Mexico. It is the most vivid, beautiful thing they have ever seen.

Over his shoulder, just after 9:00 A.M., Ceary Johnson can hear Roach screaming. Roach's voice is rising out of the babble. Roach wonders if Mikeska can even hear him, whether Mikeska wants to hear him. Overhead, two airplanes are banking and turning into the deepest part of the chemical fog. Roach is still shouting:

Move that damn ship.

William Francis Roach had always felt that wildly, wickedly disparate things and people improbably converged on Texas City.

Drag the ship into the bay right now. Move the ship into open water.

Roach, like the hundreds of people who have crowded near the ship, can hardly see. The seventy-two-hundred-ton ship looks as if it's gasping for breath. The deck planks are bending horribly, and long, rusted nails are shooting straight up into the air. Suddenly there is a long, fast series of sharp, cracking noises—like machine-gun blasts—coming from belowdecks. Roach is begging:

Please move the damn ship.

He can see that the Gulf of Mexico is bubbling, and it's as if it's from his delirious dreams. The sea is boiling.

The People

Julio Luna, Longshoreman

THE HUMIDITY is as thick as grease, and the rank smell of the factories intensifies as Luna's car pitches and rolls over the roads where the paving crews have smoothed out tons of shells dredged from the bay.

He's twenty-six, and he had seen a few things when he went to war.

Things that had made him wonder about the way the government works. Yesterday's paper had stories about the anniversary of the sinking of the *Titanic*. Today's paper says that President Truman's trusted confidant Bernard Baruch is talking about how important it is to send American aid to foreign nations. Truman's man put it this way:

"Make no mistake about it, we are in the midst of a cold war."

Luna can see that the biggest man in Texas City, Mike Mikeska, is already on the waterfront. He's talking through the cigar smoke being blown about by his vice president Walter Sandberg. The black longshoreman gangs are already there, including the one run by Ceary Johnson, his body so rippled from carrying so much cargo that it looks as if you can bounce an Indian-head penny off his arm and it would come right back to you.

Luna, lean and angular-faced, isn't a union man, but he's thinking about signing up. He likes being on the docks. It reminds him of the only time he felt even close to being really free. Luna had signed up for the navy in 1942, served as an Able Seaman on the Liberty ship U.S.S.

Starr King (named for a patriotic preacher who headed off a move to make California its own republic), and was onboard when a ruthlessly efficient Japanese submarine, the I-21, fired four torpedoes at it. The *King* had been steaming from Sydney to New Caledonia in the South Pacific. It was February 9, 1943, and with some mates, Luna watched his ship go down while he was floating in a lifeboat, bobbing on water so clear you could see the sharks as if they were in a gigantic aquarium. An Australian destroyer, the H.M.A.S. *Warramonga*, rescued the crew. As soon as he saw the Aussie ship, Luna knew he would never forget that name. He and the other men were transported back to Sydney and put up in the Grand, the fanciest hotel he had ever seen. Along with his mates, Luna was interrogated for hours by U.S. Naval Intelligence officers who thought one of the sailors had caused the ship to be sabotaged:

Did you see anybody smoke? Did anybody turn on lights when they weren't supposed to? Do you think a glowing cigarette made your ship an easy target? What were you doing that exact moment you were hit?

Luna sailed back out, onto different Liberty ships, and finally wound up as a gunner's mate on the U.S.S. *Hornet*. He went to sea again, but he was always bitter about the way he had been grilled by the investigators.

"You'd think I had sunk the damn ship. Goddamn, you think it was me?"

After the war, he returned to Texas, and there were still Mexican men living in old abandoned railcars and hitching rides into Texas City to put their names on the work lists. Luna and his family had grown up in one of those ancient railroad cars on the outskirts of town, out near the little area they call Arcadia, out where there used to be a dairy. Luna slipped right back into the rhythm of a Mexican American workingman on the Texas Gulf Coast. He knew he wasn't going to work inside the refineries or the plants, unless he got lucky and they handed him a broom. Instead, every morning, he headed into Texas City and hoped to grab that work on the waterfront.

This morning, Luna listens to the lead stevedore on the dock. Luna's been assigned, again, to the *Grandcamp*. The old ship is docked prow out to sea. It will make three straight days that he's worked on her. Monday he was down in the No. 2 hold, laying the one-hundred-pound paper bags of ammonium nitrate on top of each other. Tuesday, he had finished up in the No. 2 hold, putting the last of the 32,092 bags inside that part of the ship. He moved on to the No. 4 hold and helped load a few thousand more bags until the rain began to press down. The work whistle blew; someone shouted for him to knock off for the rest of the day. There were at least another twelve thousand bags to be loaded.

Now he is back onboard the *Grandcamp*, and he is staring at the No. 4 hold.

Luna and his crew spend ten minutes wrestling to remove the tarp covering the hatch. Ceary Johnson and his crew wait on the dock, ready for the signal to load the bags of ammonium nitrate from Warehouse O onto one-inch-thick wooden pallets. The bags will be lowered by winches into the holds, where Luna and his gang can get to work.

Luna doffs his shirt and clambers into the superheated belly of the old freighter with his buddy Bill Thompson. Luna's been on Liberty ships before. He walks through the dim light to the offshore side of the vessel, stares at the big bags of ammonium nitrate, and begins wrestling them off wooden pallets and into piles. Sweat cascading down his face, Luna looks across the hold to the side of the ship that's closest to the dock.

He smells something. Like paper burning.

What the hell?

Luna doesn't see anything. The smell is definitely there. He calls up from the bottom of the hatch:

"I smell paper burning. Is there a Frenchman burning any paper up there anywhere?"

Now there are trails of smoke along the starboard wall, at the middle point of the hold. They look, for a few seconds, as if they are actually slithering off the deck and down the hatch and into the hold.

From down in the hole, Luna can hear people on deck hollering, and he has no damn idea what the hell is going on. Luna yells again. He turns to his buddy Thompson.

"There's a fire in the hole."

Finally, Luna's gang boss is lowering a pair of one-gallon jugs of water into the hold. Luna grabs one and, waving the smoke from his face, begins scrambling over the bags. He's on top of the stack that is nearest the inside hull of the ship. Kneeling down, the jug of water sloshing alongside, he hoists the first warm one-hundred-pound bag to the side. He lies on his stomach, the heat pressing into his belly, and reaches deeper into the stack, pulling more bags, each one of them hotter than the last. Now it looks as if the smoke is actually coming from somewhere four, five, maybe seven bags deep. The damn smoke has been creeping up and out of the hold, not down and in. There is a small glow, something shining, deep in the pile of bags. It is like coal smoldering.

Luna tips the jug and lets the water splash at the smoke.

That isn't doing a damn thing.

It's hotter than ever inside the hold; he is completely coated in sweat. The light is weak and the smoke is almost alive, reaching out into the corners of the claustrophobic hold. Luna watches as two French sailors lower themselves down. They are carrying two five-gallon fire extinguishers. They play the fire retardant onto the stacks, and the glow goes away. Five seconds later it reappears. It is a yard long. Then there is a small flame, about ten inches high. The French sailors aim their extinguishers, and the fire instantly dies.

Then, as if it is playing cat and mouse, another little flame pokes up along another section of the starboard wall. It is put out, and then another ten-inch tongue of fire shoots up farther down the row of bags.

Luna stares at the dancing flames.

This is no good.

Luna is climbing out, crawling onto the deck, and looking down at the dock. Mike Mikeska is there. Father Bill Roach. Fire Chief Henry Baumgartner. Behind them, he can see Ceary Johnson mingling with his men. There are children out there on the seawall now, leaning against their bicycles and shoving one another for a better view.

As Luna walks away from hold No. 4, the French captain, Charles de Guillebon, is ordering the ship's emergency whistle to sound. At 8:20 A.M. there are three short blasts and one long blast. De Guillebon is staring at the way the smoke is beginning to bellow, as if someone has fired up a chimney. De Guillebon talks in low tones with his officers. Several French sailors fan out in a line around the opening to the holds. De Guillebon issues an order to begin sealing the hatch.

For a second the green-orange smoke seems to slink back inside the ship like an eel retreating into a wall of coral. Now de Guillebon is huddling with his engineers, and they are talking about opening up the valves that can fill the superheated No. 4 hold with pressurized steam. Flooding the hold with water will certainly ruin the cargo. Steam will be better. Steam will save the cargo.

Over the last week, de Guillebon has been writing what will be his final letters to his family. So have his sailors. In his last letter to his wife, mailed two days ago from Texas City, de Guillebon tried to soothe her. He knew what she was thinking. He could almost hear the sorrow in her voice rising like a painful and predictable flood. Whenever he came home, she stared at him with the knowledge that he was going to be leaving again. He was always going to go back out to sea again . . . and the cycle would go on and on as it always had.

"I know the solitude," the captain wrote to his wife. "I don't see myself making more than one more voyage after this one."

Now he wants to smother this damned fire. Save the cargo, smother the fire, get out of Texas City, and just go home. The sailors begin to unroll a bleached-looking cloth tarp and drape it over the smoking hatch. They grunt and grip timbers and throw them on top of the tarp. They want to starve the fire. Deny it oxygen. They want to watch it sputter and die.

While some of the sailors work on the No. 4 hold, others are jogging to the No. 5 hold. There are those sixteen cases of small ammunition in No. 5 that somehow were never unloaded in Venezuela.

Unload those damn crates.

Julio Luna watches the sailors running for No. 5. He volunteers to go into No. 5 to help.

He drops down into the hold, sees the wooden crates, and begins to shove hard against them. His back and shoulder are into it, but they aren't budging, they can't be moved. Some of them are tipped over. And now there are tongues of that smoke licking their way into the No. 5 hold.

Luna pulls himself out of the hold and walks briskly to the gang-

CITY ON FIRE   87

plank. Now he's onto the apron around the wharf, moving with a small line of other quickly moving longshoremen. Jimmie Fagg, one of the members of his gang, yells out that if the fire department shows up they're going hose down the *Grandcamp* with water sucked out of the bay. And that won't be good at all.

Who the hell knows what chemicals are in that water?

Luna's not looking back as his work boots race over the broken shells and the oil-caked mud. It's just after 9:00 A.M.

Mike Mikeska and Walter Sandberg, the Company Men

He backs out of the driveway that morning while his daughter makes a fresh pot of coffee and mumbles something about going to another Garden Club meeting.

Mikeska turns left, toward the docks, and finds his usual parking space alongside the Texas Terminal Railway offices. The building is 750 feet southwest of Pier O. Mikeska sees Pete Suderman, the guy who represents the different unionized longshoreman's gangs that work the waterfront.

Suderman has his sign-up sheets, the ones that tell him what gangs are available, who got the last bit of work, who is entitled to some extra hours.

The rain on Sunday had slowed things down. There are still six hundred to eight hundred tons of ammonium nitrate to load. Suderman and Mikeska want the best crews, the most experienced crews, to load the stuff, especially if the French freighter is going to sail out tonight. Mikeska sees Ceary Johnson, Julio Luna, and the pockets of anxious longshoremen. French sailors are on the deck of the *Grandcamp*.

Some union representatives are checking their watches, making notes on their ledgers, seeing who is on the clock and who isn't.

Mikeska turns and walks toward his office. He and his aide, Walter Sandberg, have the same feeling.

The damn unions have a stranglehold on things.

Sandberg is even more blunt.

"You don't know those longshoremen. They don't like to work that fast."

Mikeska had graduated from Texas A&M University, one of the most rigid schools in the South, and he had meshed perfectly with the colonel—Henry Moore, the man who ran The Company and Texas City for forty years. Mikeska had studied civil engineering, and Moore liked his ramrod way of doing things, his addiction to making the trains run on time. Moore had always believed that machines could move in concert, that ever-growing numbers of ships and trains could be forced into orchestration. That's what the colonel sold when he was out hawking Texas City.

The place moved on time; there were no roadblocks; it worked like an oiled engine to receive cargo, oil, and chemicals—and then to warehouse them and ship them anywhere in the world. For years, as Mikeska climbed the ladder at The Company, there really was no one else to whom the colonel was going to turn.

Mikeska was the perfect choice. Everyone knew it; everyone could see that Mike Mikeska was as demanding, dominating, and inflexible as the colonel himself.

More than a few people in town half expected to find Mikeska out there on the waterfront during the bad hurricane in '43 and doing the same thing that Moore had done when the last great storm tried to lay claim to Texas City in 1915: back then Moore had stepped to his doorway, the howling winds practically blowing his beard off, and he began shouting into the dark rain, demanding, ordering the damned hurricane to get the hell out Texas City.

Now his heirs, Mikeska and Sandberg, are overseeing the eleventh-largest port in the United States, and more than twenty-five hundred ships are passing through the docks every year. In Mikeska's office, the secretary Joyce Vogg is hunched over an adding machine. E. J. Katzmark, cashier for The Company, is waiting his turn over Vogg's shoulder.

In the crowded but pleasantly worn office are train schedules, freight tariff charts, lists of shipping agents, maps of the Gulf of Mexico, bills of lading from the thousands of shipments that have arrived in Texas City over the last few decades, and something called "the jumbo book"—the book that contains a record of all the railcars passing along the Texas City tracks. Also in the office are the paperwork and checks from the French Supply Council, the New York–based agency that is purchasing and handling the transfer of the ammonium nitrate to the French vessel.

The two office workers are whispering about the smoke they have both just seen out the window.

Vogg has a question: *"You think we could get hurt if that ship ever exploded?"*

Katzmark replies: *"We would never know what hit us."*

In the office, Mikeska stops to talk to Sandberg. They have the same relationship that Mikeska and the colonel once had. Mikeska had watched Sandberg, the little kid who had started working for The Company the same year World War I ended. He had been there since 1917, starting as a storeroom boy. He did it all. Purchasing department. Accounting department. Operations. Now the silver plate on his desk said vice president and auditor. Sandberg had finished only two years of high school, but Mikeska liked the way he had single-mindedly spent the last thirty years with The Company. Sandberg was just as gruff, just as intense as Moore and Mikeska could have hoped for.

Sandberg goes into his office. A clerk sticks his head in.

It is 8:20 A.M., and the French captain on the *Grandcamp* has just sounded the ship's whistle.

Without any hint of alarm the clerk says: *There is a fire down at the French ship.*

Sandberg is not concerned. He sifts through the audits on his desk. Five minutes after the clerk stopped by, he hears the siren begin to sound from inside the fire-pump warehouse owned by Texas City Terminal Railway. As he looks up, he can hear Mikeska ordering his personal assistant to get on the phone to the tugboat companies in Galveston.

Even though there is a national phone strike, a handful of local phone operators are still on duty to handle only the most pressing emergency calls—and sometimes the phone operator simply decides what is an emergency and what isn't. When someone with Mikeska calls, that call always goes through.

We need two tugboats to report to Texas City.

It is 8:30 A.M.

As Sandberg strains to listen to Mikeska, through the insistent sound of the siren, he swivels in his wooden chair and looks out the window facing the direction of Pier O. Orange spirals are twisting into the sky. As Sandberg studies the action, Mikeska barges out the door. Sandberg reaches for his coat and follows him on the five-minute walk through the warehouses and down to the apron of Warehouse O, immediately astern of the *Grandcamp*. He can still hear Mikeska's aide on the phone, calling to Galveston tugboat operators to ship those two tugs to Texas City.

Ahead of him, Mikeska is already deep into it with Pete Suderman, the guy overseeing longshoreman operations. They are standing two hundred feet away from the *Grandcamp*. Suderman doesn't seem alarmed. Neither does Mikeska. Sandberg joins them and listens to Suderman: *There's a fire in the No. 4 hold. The French captain wants to batten down the hatches and pump steam in there.*

Sandberg leans in and jokes to Suderman: *"What are you trying to do, burn up our port?"*

Months later, Sandberg will remember the moment: *"We stood around and discussed things in general. We were not too concerned about a fire onboard a ship. We probably talked about more things that had no connection with the vessel than about the vessel itself . . . the fire concerned us least of all."*

As Sandberg, Mikeska, and Suderman huddle for the next thirty minutes, several people stop to chat, hang out alongside, or lean in and listen. City commissioner Bill Voiles is there. Republic Oil refinery public relations man Basil Stewart is there.

At 8:40 A.M., Suderman breaks away to yell up to the longshoremen whom he can see on deck. He wants them to clear off the ship, to knock off for a while. There's no more loading for the time being. Ceary Johnson, the best-known black longshoreman in Texas City, is stepping down from the *Grandcamp*. Julio Luna is following Johnson off the ship. Several members of the French crew are clambering down from the ship, some of them pulling on shirts and pants as they join the longshoremen headed dockside.

By 9:00 A.M., Sandberg is thinking about the tugboats that Mikeska ordered. He decides to head back to the office to make some calls to check on their progress. Tugboats usually dock in Texas City or just outside the harbor, like seaside taxis waiting for the next call. This morning there are none.

As he walks back, he can see the growing crowd watching the brilliant smoke. It is just after 9:00 A.M.

Florencio Jasso, Merchant Seaman

A buddy of Julio Luna's named Florencio Jasso is lucky . . . if you can call it that. He's joined the union, and they're tipping him off to cargo ships that need crews for short jaunts at sea. At 7:30 A.M., he's checking his watch. He needs to be in Galveston before too long. He needs to report to the merchant seamen's office there to see if any work is available.

Jasso is another World War II vet, a ball-turret gunner who spent several months locked inside a see-through bubble affixed to the underbelly of a B-17 Flying Fortress bomber from the 20th Wing. Jasso was locked in a gunner's seat almost guaranteed to put him first in line to be killed by antiaircraft fire, or first in line to be killed if the plane crashed.

Jasso had been all over Europe. His plane had been hit by flak. He had been at twenty-seven thousand feet when the pilot lost control and it looked as if the bomber was going to belly flop; Jasso had been left wondering if he was destined to be another turret gunner who absolutely knew there was no way in hell he was going to survive inside that glass cage, who knew that he had to just sit his ass still, watching the ground zoom up and somberly reciting his last wishes to the rest of the crew.

Like Luna, Jasso survived. He came back home to Texas City, and he also thought a lot about the fact that Mexican Americans had laid the tracks and built the buildings but weren't being hired inside the refineries. He was looking for work as a merchant seaman, something that a Mexican man could do without too much prejudice. Back home after the war, he moved with his parents into a tiny home in the middle of a small island of shacks owned by The Company, right in the middle of refinery row and two blocks from the north slip.

His old man had come up from Sierra Nuevo Leon in northern Mexico, looking for work, and he was met on the American side of the Rio Grande by men looking to hire cheap labor. They were all over the border—bored-looking men holding watch fobs, squinting against the sun, and resting two-toned shoes on the running boards of their cars. They were the recruiting agents for the oil refineries and chemical companies. They had been dispatched down to the Texas-Mexico border to steer workers to the refineries, docks, and plants along the Gulf Coast.

The men sometimes had a joke.

"We're looking for Mexican dragons."

People would shake their heads.

Dragons?

"Yeah, you know, we need some Mexicans to drag some shovels, drag the dirt, drag the shit from one place to another. Mexican draggin's. Get it?"

His old man is nicknamed Shorty, and he has finally wound up working for The Company as an "oiler"—a man with an oil can who crawls all over the boxcars making sure the switches are properly greased. His mother is a saint, someone who one way or another keeps the family together. People call her son Flo, short for Florencio, and like Curtis Trahan, like Julio Luna, like a lot of people, Flo looked different when he came back home from the war and moved into one of The Company's old shacks. He looked different, but things in Texas City didn't seem all that different to him.

This morning, he needs to be in Galveston. He has joined up with the National Maritime Union, and he has been checking into the regional headquarters, in Galveston, to learn which ships need crew members. Flo says good-bye to his mother, and he walks past Forbes Drug and Davidson's Grocery. He moves north on Sixth Street and steps inside the Rainbow Diner. He orders coffee and checks the time. The 8:30 bus will take him over the causeway—the causeway his old man helped to build—and into Galveston.

The diner door opens, and an old friend, a longshoreman, spots him. His friend has a camera and a car.

C'mon, there's a ship on fire down at the north slip. I'll give you a ride.

Flo thinks about it for a second and then hops in the guy's car. Up ahead, he can see the smoke set in the perfect sky, moving like orange waves breaking on a bright blue shore.

Forrest Walker Jr., the High School Senior

Forrest loves his old man, though they are not especially close. His father doesn't have any bad habits, none that his son knows about. He doesn't drink, smoke, curse . . . except for the one time the old man plowed his truck into something and shouted, "Oh shit!" His father was never baptized because he was raised as a Quaker, but, in Texas City, he joined the Methodist church and the family went as often as they had services. At the church, and sometimes in private functions, people liked his father's singing and his pleasant baritone voice.

Forrest is skinny, at six feet one inch and 135 pounds, and he feels that his old man might be ashamed of him. He also feels that his father is an outright bigot. Once, during an evening sermon at the Methodist church, the minister talked a little about racial equality. Afterward, Forrest watched as his old man confronted the minister and told him he disagreed. Still, Forrest sometimes sees his father mingling politely with black people. Sometimes Forrest thinks his father talks one way and acts another.

His father is strict, straight. He is a provider, good to his wife and family.

Four years ago, he listened as his father, who was barely scraping by with his job at a fruit-and-vegetable warehouse in Kansas, announced that the family was loading the car and moving to Texas. Things were happening in Texas: factories were opening up overnight to feed the war effort, and the docks were bustling.

Forrest's old man found work on the dry docks in Galveston and almost immediately heard about openings a dozen miles up the bay at the Monsanto plant in Texas City. The government had handed the abandoned Texas City sugar refinery to Monsanto so the chemical giant could run a round-the-clock war plant churning out synthetic rubber. The place had its own power plant that needed to be stroked and coaxed back to life. Forrest Sr. applied for a job as a boilermaker, joined the American Federation of Labor (AFL), and the family moved up to Texas City in the middle of the war.

Forrest Jr. was walking down Sixth Street one day when Curtis Trahan loped out of his insurance office: *Do you want a job?*

Forrest simply said: *"What kind?"*

Trahan said that he was going to run for mayor, and he needed someone to hand out some flyers. Forrest walked around downtown for an hour, offering one to anyone who would take it. When he finished, Trahan said he'd pay him $10 a week to be an office and errand boy. Forrest thought, as did plenty of other people, that Curtis Trahan could actually run for higher office.

This Wednesday, Forrest has just finished dressing after his physical education class. Forrest, Buddy Rich, and other pals are walking back to the main high school building. Forrest is preoccupied, almost annoyed about that school play. Forrest tried out, hoping to be some human wallpaper in some tiny scene, just so he could get his name in the yearbook. They wound up giving him the lead as "Old Doc"—and he hadn't learned any lines and didn't even know what the play was about. It all looked like a giant disaster in the making.

Just before they push open the door to the main school building, the three friends see the orange smoke coming from somewhere down by the waterfront. Buddy wants to go check it out. Forrest never plays hooky, so he says he won't go; his friends sprint off to pile into a car.

Forrest walks down the school hall, goes into Mrs. Lash's typing class, and takes a seat next to another buddy named J. B. Meyers. Mrs. Lash is instantly all over him, waving a marked-up homework assignment under his nose and crabbing at him. *This is filled with typos. I want you to do it again.*

Forrest hates the idea, but he reaches for a clean piece of paper and starts to roll it into his typewriter.

The clock has just ticked forward to 9:12 A.M. As he glances out the window, he catches a glimpse of what looks like a streak of day lightning. Maybe a squall is coming, something big exploding out over the Gulf of Mexico.

John Hill, the Chemical Engineer

The room won't stop spinning.

He's twenty-eight and he never misses work. Never. Now it's 8:30 A.M., and he's just looking to lie down.

He had gotten up at the usual time, 6:00 A.M., but his stomach felt like hell and the room was already whirling. Still, he had kissed his Margie good-bye, patted their two-year-old boy, Jae, on the head, and headed out the driveway of the family home on Seventeenth Avenue North.

At the Union Carbide plant, a few hundred yards farther away from the waterfront than Monsanto, people say Hill is a climber. He doesn't mind the description. He always felt as if he had been born to hustle, on the outskirts of San Antonio in a family with fourteen kids. He always figured that all that Depression-era jockeying around the dinner table had made him damn competitive. Thank God his father had run a grocery store in addition to his job running the cottonseed oil mill.

Hill went on to attend the University of Texas, and he got a good look and the offer of a minor league contract from the New York Yankees. He loved smashing the ball to right field, but he didn't like the money the Yankee scout was offering, so he followed up on the word that Carbide was hiring in Texas City. He arrived in the summer of 1941 with one suitcase, and he moved into a garage apartment with two other young Carbide guys.

Three months later Hill took the leap.

I'm going off the springboard.

He finally married that hometown girl, and he and Margaret—everyone called her Margie—found a fourteen-hundred-square-foot house way up on the north side. Hill began making his name at Union Carbide, working different shifts and doing the nuts-and-bolts work of refining petroleum and petrochemical products—cracking methane and propane, processing sulfuric acid, removing hydrogen sulfide and acetylene. Margie sometimes wondered if it was dangerous working deep inside the petrochemical maze, with sulfuric acid and all the other toxic compounds. Hill calmed her and said it wasn't really dangerous at all:

You go where the money is. You go where the job is. It's only dangerous if you're foolish. The danger isn't in the process. It's not the equipment. It's the people.

Just forty-five days ago, Hill had been given a promotion. Now he's heading up one of the gas separation units in the No. 1 plant. Sometimes he thought about how far he had come: he was the only one of those fourteen kids who graduated from college. He took his summa cum laude degree to a place called Texas City, and now everyone knows his name at the Lions Club meetings at Lucus's Café on Sixth Street. Unlike a lot of people at Union Carbide, Hill believes in fitting in, hanging out with anyone in Texas City—he's that way, anyway, but he think it's probably real good for business.

He's become more than a fixture at the Lions Club. In a town dotted with optimistic transients, the club is a rock-steady way station for men looking for business partners, gossip, and the camaraderie they'd felt on the front lines of World War II.

Hill had met Curtis Trahan at the Lions Club. Now he and Margie have been playing bridge with Trahan and his wife, Edna. Curtis may be a brother Lion, but he's a lousy bridge player. He lacks the killer instinct. And, besides that, he's more than just a union sympathizer; he's way too liberal. Hill likes to remind him: *You're too damned liberal. I'm pure conservative.*

It takes all types, Hill always figured. He had been in town for only a few years, but he felt as if he had it pretty well figured out. This is how he framed it to Margie: *Texas City is the greatest example of the theory that whenever a vacuum exists, something will rush in to fill it.*

This morning, still dizzy, he had crawled in to work at 7:30.

If you take the man's money, you oughta show up for work.

By 8:00 A.M., it's worse than when he first stepped out of bed. He finds his boss, Bob Hieronymous. "I need to go home and get to bed," says Hill. Bob tells him to get up and go. At 8:30, he's pulling back into his driveway.

He has seen the brownish orange plume in the sky, but his head is cloudy, he can hardly focus, and he just wants to be inside. Margie has taken Jae to visit some church friends, to their house somewhere near to the port area. Hill falls back onto his bed and closes his eyes, and the room begins to quiver at 9:12 A.M.

Elizabeth Dalehite, the Sea Captain's Wife

Hands on the steering wheel, the forty-six-year-old Elizabeth looks over at her forty-seven-year-old Henry as they take the fork that will deliver them to the Texas City waterfront.

It's 8:45 A.M.; he's bundled up, half-asleep, his thick body poking out from under an old blanket that he likes to carry with him when they're out driving at night or the early morning. Though they had tried to have another child ten years ago, the baby died eighteen days after she was born. Now they've been married twenty-three years, and she's been through it all with him . . . tying the boat lines down when the '43 hurricane came in, towing boats from Houston to Corpus Christi . . . rescuing from freighters panicky sailors who couldn't figure out how to negotiate their way around the sandbars off Galveston Island . . . making runs to Baytown, Aransas Pass, and Port Arthur.

She sees Henry look up groggily at her, the woman he sometimes calls Mama Lu.

It's one of those silly family things. She's Mama Lu, because when she came home from the hospital her father had said she was a lulu. Three months later, she was floating out of her house on a wooden refrigerator, thus becoming one of the few fortunate survivors of the Great Storm of 1900—the one that covered all of Galveston Island in water and killed as many as eight thousand people. She was a lulu then, and now she's Mama Lu, living with her boisterous sea captain husband—who had also ridden out that monster storm—in a two-story house above Offat's Bayou set on tall pilings made from eight-by-eight-inch timbers.

Henry is always working on the gritty, hard, humid docks, and their two children, Betty and Henry Jr., hardly see their father. He's got contracts with Seatrain to guide their ships all over Galveston Bay and the Gulf of Mexico. He's got an excursion-and-pleasure boat called *The Galvez* that has a few illegal slot machines onboard and that runs gamblers, lovers, and politicians on jaunts around Bolivar Peninsula. He's got three small fishing boats that he's named *The Laura Foster, The Josephine*, and *The Carol Anne*. One of his thirty-nine-foot launches is named *The Betty,* in honor of his first child. There are times when Henry gets a call and leaves in the middle of the night. His wife can hear him pulling on his clothes, and she gets up, too, to make him something to eat, to give him something warm to drink. Off he'll go, picking up his mates; then he'll be gone for days, guiding oil and gas barges through the Intracoastal Canals in Louisiana and then into the Mississippi River.

Henry has another unofficial job that he's created for himself. After he's finished guiding another oceangoing freighter into port, he's up late entertaining the sailors and the captains. Only a few inches over five feet tall, Henry's built like a fire hydrant; he's a funny guy, a tough guy, somebody everybody knows in almost all the waterfront bars. After making the rounds, he knows full well that Mama Lu will have a huge home-cooked meal ready for him and his new friends. She doesn't mind. She really doesn't. She loves to cook, and she gets angry if you don't eat and ask for seconds and thirds. And she loves to hear the sagas being spun by the old salts. So do the kids, their ears pleasantly burning as they sneak around in the shadows, trying to listen to their exotic visitors, working hard to pick up snatches of their mad seafaring stories.

People love being at the Dalehite house; they love everything about it, including walking out on the oyster reef in the bayou, gathering up buckets of the shellfish, and watching them smoke on an open fire. The visiting sea captains hardly want to leave. Henry and Elizabeth Dalehite are in love with what they're doing. They're not fancy, not rich, but they're able to connect with almost anybody else who's doing what they've already done for three decades . . . carving a life out of the sea. Henry is down-home and wide open and funny as hell.

You know why they call it Offat's Bayou? 'Cause that's where the damn

railcar used to stop, and those old-timers used to say "let me off at the
bayou."

Off at.

Offat's.

Get it?

And this morning, Henry is bouncing in and out of sleep on the passenger side of their car.

Elizabeth . . . Mama Lu . . . is used to it. She does the driving, taking him up to Houston, to Texas City, to anyplace along the Texas coast where people have called and asked for Henry Dalehite to bring their ship to port. He knows the tides, the shoals, and what's behind the curling fog on Galveston Bay.

This Wednesday morning, of course, the damned phones aren't working. There is that national phone strike, and it's almost impossible to get a call through unless you've got a relative or a best friend who happens to be a telephone operator. Communication is completely screwed up. Only "essential" calls are allowed to go through, and the exact meaning of "essential" is up for grabs.

Henry has been waiting to hear about the *New Yorker*, a Seatrain ship that will need help being piloted from six miles offshore, out near the bar in the Gulf of Mexico. It's due in Texas waters sometime tomorrow, but Henry hasn't heard a damned thing about exactly when. Seatrain can't call him, and he can't call Seatrain. Elizabeth and Henry have been up since 2:30 A.M., trying to convince the phone operators that it really is essential, that they really need to be patched through.

The operator, a man, told them shipping didn't sound essential, which is about the worst thing you could say to a Dalehite.

Elizabeth had to laugh.

Henry told the man what he thought of him.

Exasperated, Henry finally asked Elizabeth to drive him over there, down to Seatrain's little office on the Texas City port. When they left the house, it was still dark. Henry wrapped himself up in his favorite traveling blanket and watched her drive. Suddenly he suggested that they stop and see their newborn grandchild, their first grandchild. She is their daughter Betty's baby, just three months old.

Elizabeth was reluctant: *I don't want to wake anybody up.*

Henry insists. Their son, Henry Jr., is off at college, in Austin at the University of Texas. They lost their chance to have another baby ten years ago. Betty's baby is special.

Let's run in and see the baby. They won't mind. They'll all be up.

Of course, he is right. Betty, her husband, and the baby are all up. They stay for a while, watching everyone eat breakfast. Finally, they continue toward the waterfront. Elizabeth turns down the familiar narrowing road that puts her near the Monsanto plant and fifty yards away from the crane the Seatrain company uses to load boxcars from the railroad tracks and onto its freighters. Elizabeth finds a place alongside a chain-link fence surrounding the Monsanto plant.

She looks over at her husband. Henry has fallen asleep again. She gently touches him on the shoulder. *"Look, there's a fire over on that ship."*

It's 9:10 A.M. Henry looks up and stares for a few seconds at the commotion on Pier O. He opens the car door, steps outside, and turns back to Elizabeth.

"C'mon."

She is reaching down for the little statue of the Blessed Mary that she carries in her purse. Elizabeth has a habit of saying her morning prayers when she's waiting for Henry to come back from the sea or when she's waiting in the car while he cuts his deals. She likes to stay in the car, watching the sun come up, starting the day with her prayers. She is bone-weary and slipping back against her car seat. She smiles up at Henry. *Honey, I'm too tired. You go ahead.*

She begins to whisper her prayers, her hands wrapped around the statue. He walks away. Henry turns back for a second, and she sees him waving at her. Then, there is a blinding light.

Rev. F. M. Johnson, the Preacher

Bill Roach loves to visit him. The Baptist preacher has fat clumps of okra and other greens coming, almost magically, from the salty soil. They seem to defy the constant chemical clouds that blow in over The

Bottom when the winds shift and come from out of the Gulf of Mexico. Sometimes, when the flailing memories of the tropical storms turn into slow, small, warm rains, it can seem as if the drops are a different, darker color as they catch the thick haze and settle on The Bottom.

Johnson has been in Texas City for only a year, but he has been drawing more and more people to his church, sandwiched between the Methodist congregation and Barbour's Chapel. Johnson is young, handsome, and prone to wearing tailored suits and sunglasses. He is also unafraid—utterly, completely unafraid. When people come to hear his sermons on Sunday, they leave saying that he seems to march to his own drummer.

At Norris's Café on Sixth Street South, people are talking about him. The consensus is that the new preacher is someone that Texas City will hear a lot about before too long. And, besides, the new Catholic priest in Texas City, Bill Roach, has already singled him out as someone worth knowing.

Before Johnson had taken over the reins of First Baptist, he cut his teeth in hidden places in Texas where the Ku Klux Klan was especially virulent. He had been born in Maynard, out near Sam Houston Forest, deep in a part of Texas where almost everyone could tell a story about a place in the woods where men were hanged. He was raised in Conroe, near the San Jacinto River and not far from the town of Cut And Shoot. In the 1920s, when Johnson was a boy, the hangings seemed to happen every week. The long stretch of East Texas, running from the Red River to Texas City, is a place where people routinely disappeared. People knew the stories; they knew that although things were not as open, they were just as intense right into the 1940s. In 1941, in Pittsburg, Texas, a black janitor was dragged from jail by a crowd of two hundred people and castrated. For two days in 1943, in one of the worst race riots in American history, crowds as large as four thousand roamed the streets of Beaumont, Texas, with guns, axes, and hammers, looking for niggers to kill and maim. The KKK had announced that it was hoping to have twenty thousand of its members converge on Texas for its national meeting.

Now, this morning, there is a teenager named Buster Bridges from the neighborhood at the door of his low-ceilinged house in The Bottom. Johnson is in the shower. He ducks his head outside to see what the young man wants. Bridges wants to know if Johnson wants to get his car washed. Johnson hands him the keys out a window, and the teenager pulls the car out of the garage and lines it up near a water faucet in the front yard. Bridges fills a bucket and begins to run a brush over Johnson's '41 Chevrolet.

Johnson is in a hurry. He has a Baptist board meeting in Houston later in the day. As Bridges works on his car, Johnson quickly dresses.

Suddenly Bridges is knocking on the door again. "Rev, let's go down to the docks; there's a ship burning."

Johnson says he has to finish getting dressed and then he has to go straight to Houston.

Bridges insists. "Well, it ain't going to take long."

Johnson decides that he has some time to spare. He'll finish getting dressed and then he'll drive down with Bridges to see the fire. Then there is a deafening roar.

Kathryne Stewart, the Mother

She never stops marveling at the fact that her children have such extraordinary blond hair. She has brown hair; her husband, Basil, has dark hair. Their children, Marilyn and Stu, look almost like porcelain dolls with their rosy complexions and bright, cascading manes of that impossibly blond hair.

Kathryne Stewart has six-year-old Marilyn and five-year-old Stu in the car. They've been in Texas City ten months now, and Kathryne always drives Marilyn to school. They're on their way to Danforth Elementary. Basil has already left for his job as director of industrial relations for Republic Oil refinery. He's a lawyer and the public relations spokesman, and people say he is tailor-made for the job, a perfect ambassador moving between the community and Republic's corporate interests. People trust him, even if they might be suspicious of the company for which he works. He is the messenger for the company, not the one who makes its policies, and people like Basil Stewart because he is a consummate "joiner"—someone who has to be in on every play, who has decided that immersing yourself in the community is always a win-win situation.

At the front door that morning he lingers while staring down and giving Kathryne a kiss. They are both forty. He is six feet four inches, 240 pounds, with narrow gray eyes, thin lips, and a mound of curly black hair that he parts on the left side of his head. He says he is going to run up to Houston with Mayor Curtis Trahan—they've become almost unlikely friends through the Lions Club, and now Trahan is going to help him in a bid to become a regional director for the club.

Basil says he might not be home at the usual hour. The children might be disappointed. He has a regimen: when he knocks off work and sets foot in the house, the children are his priority. Anything they want to do, they can do—and he always tags along. Maybe their favorite thing is to go for a car ride to the park or, better, to a place where they can get a pony ride. Back home, after dinner, they crawl over the back of his easy chair, running their fingers through his hair, trying to see what he is reading.

Kathryne tells friends and other members of the family that when she meets other men, she measures them against Basil. Are they as intelligent? As patient? As attentive? She has been planning their tenth-anniversary celebration. It's coming up in July, and she has been thinking about how she and Basil waited until they were each thirty before they got married. She had been a teacher; now she stays at home with the children, and she has made complete peace with her decision. It is, for her, the absolutely right thing to do. And Texas City seems like the right fit for the family.

They had lived in another industrial town in Texas, Galena Park, before coming to Texas City. They hated to leave. Back then the Stewart family was probably the most active one in town: Basil had been mayor, he had served on the school board, he had been president of the Lions Club and the Masons. Kathryne was there, every step of the way.

But Republic made Basil a good offer, and everyone could see that Texas City was rocketing ahead. Now, in Texas City, they are beginning to immerse themselves in the community. Just yesterday, Basil went to a photo studio to have some nice portraits taken that might help him move up the hierarchy in the Lions Club—pictures that he can take with him on his trip to Houston with Curtis Trahan.

This morning, when he leaves the house, he makes sure to be wearing his past president's Lions Club button, a Shriners tie clip, and a company watch.

Now Kathryne and the children are at the front of the school.

It's just after 8:30 A.M. and Marilyn, dressed in a gingham dress, bounds from the car; her mother calls her back for the usual good-bye kiss. Then Kathryne stares at her daughter disappearing into the pack of children. Kathryne feels blessed. Her children are smart and beautiful. Basil, she thinks, is simply brilliant and handsome. And, she tells people, he is a ridiculously generous soul—someone who makes time, after work, to help countless people with the legal paperwork for their wills, estates, adoptions, divorces, and jobs.

Kathryne has a 9:30 A.M. committee meeting at the PTA.

In the car, five-year-old Stu is banging on the seat: *Look, Mama, ain't it funny-looking smoke?*

Kathryne looks out the car window. She corrects her son's grammar: *Isn't it funny-looking smoke.*

She looks at her watch. She has a few minutes to spare. Why not indulge the boy? Why not stop and look at the smoke for a minute or two? She turns south on Sixth Street and joins the growing traffic moving toward the docks. She turns left on an old shell road that leads to the waterfront. Stu races from the car, vanishing into the crowd alongside the seawall. Kathryne finally breathes easy when she finds him and they hold hands, watching the orange smoke moving in carpets up to the skies.

She squeezes Stu's hand. *It is beautiful, but we have to go now. Momma's got to go. I've got my PTA meeting.*

The boy pulls back. *I don't want to go. I want to watch the fire. Momma, please.*

Finally, he relents, and they are back in the car, making the short drive back to their house on the north side. Kathryne tells Stu to stay in the car; she's coming right back out. She walks in the house, pulling bobby pins from her hair and reaching for a comb.

That smoke was strange.

Henry Baumgartner, the Fire Chief

Harold's old man, Henry Baumgartner, had moved the family into a comfortably crowded house right behind City Hall. People see his father everywhere and his father is, almost, a town character. His old man knows the first name of every adult and child. Henry has a ruddy, jowly face, a firm handshake, and a nimble walk. Henry's wife, Christine, and his four children worship him, including their twelve-year-old Harold, who wonders if he will become a fireman like his father.

This morning the two little girls who live next door to the Baumgartners on Seventh Street rush, as usual, to their front window. It's their secret small-child treat. Patricia and Meta Kirby peer over the windowsill, giggling and waving good-bye to Henry as walks to his car and heads to work. The people who live on this stretch of Seventh have been neighbors for years. There is a bond that has set in among the Baumgartners, the Kirbys, and the Madduxes. They share everything. Doors are left open, and the children from each family are welcome to come in anytime to talk, to sit at the kitchen table and sip lemonade. If someone's pet is lost, the whole block seems to move in unison to search the bushes and the alleys.

This morning, a smiling Baumgartner knows the Kirby girls will be in place. They are like clockwork, and he has grown to worry when they are not there. He looks up and waves back to the five-year-old and two-year-old girls as they press their faces against the front window of their quiet home. Just a few days earlier Baumgartner had been reflecting on the simple rhythms of their little zone in Texas City. The Kirby girls seemed to embody the whole thing. He told his wife, Christine: "How I enjoy having the girls wave to me every morning."

As Henry steps into his car, Harold is finishing breakfast and preparing to open his schoolbooks. He wants to sit by the window and work on some homework. His parents know him to be self-disciplined, someone who can summon up the strength to resist the urge to hop on his bicycle before he has finished his work. He is sometimes very serious, someone who appears to have a level head on his shoulders. In a small town deep in the soul of Texas, he is the fire chief's son. The one who sometimes gets to ride in the trucks. The one who gets to hear all the stories firsthand.

More than a few of Harold's friends thought he was the luckiest kid in Texas City.

He watches his father drive out onto the familiar streets to the docks.

By 8:00 A.M., Baumgartner, head of the Texas City Volunteer Fire Department for twenty years, is beginning his regular job as purchasing agent for the Texas City Terminal Railway. He checks in with the clerks at one of the warehouses, about five hundred yards from the *Grandcamp*, that contains rail supplies for repairing locomotives. His office is in the same warehouse.

By 8:15 A.M., a clerk for the railway has told him that the smoke is getting bigger on the *Grandcamp*. The longshoremen and the French crew couldn't put it out with water buckets and the ship's small fire extinguishers. Of course, there is no fireboat in dock—Texas City had begged the corporations along the waterfront to help pay for the maintenance of the city's one fireboat, but the companies refused, and Texas City had been forced to sell it a few months ago.

Now Baumgartner steps outside to stare at the smoke. It is rising, and it is queer-looking. Chemical fires can sometimes send up multihued puffs. He has a bad feeling. He orders the dockside siren to sound at 8:30 A.M. Three minutes later, he is grim-faced as he shouts for Harley Bowen, the Texas City Terminal Railway foreman, to call in a general alarm.

"This is a dangerous fire, and we need all the help we can get."

At 8:35 A.M., Chief of Police Willie Ladish receives the frantic call, and by 8:37 A.M. he has flipped the switch activating the City Hall siren. Up on Seventh Street, amid the small homes bunched neatly together in the middle of the city and just behind that City Hall siren, Baumgartner's wife, Christine, knows something has taken a turn for the worse.

At Pier O, Baumgartner makes a decision. He wants a crew to run out a fifty-foot line of fire hose from the pumps inside Warehouse O. There are three pumps, each with 1,000- to 1,100-gallon-a-minute capacity, and they are tapped into an underground reservoir that holds

277,000 gallons. As Baumgartner directs his men, the first of the four Texas City fire trucks arrives. It is 8:45 A.M.

Under orders from Captain de Guillebon, ship stoker Pierre Andre and the rest of the crew have debarked. The French sailors are huddled at the end of the wharf, some of them coughing from the smoke that had engulfed them in the engine room. Six of the sailors decide to go for cigarettes at Frank's Café. They begin to walk alongside Warehouse O, heading past Father Bill Roach, Mike Mikeska, and Ceary Johnson.

For the next fifteen minutes, some of the volunteer firemen walk over from the adjacent refineries. A few drive up from their day jobs around the city. At 9:00 A.M., twenty-seven volunteers are assembled and taking orders from Baumgartner as he orchestrates their movement around the docks. He's wondering if some of the trucks can get a better angle on the smoke by driving into the warehouse adjacent to the docks—and then having someone open the dockside doors of the warehouse so the fire trucks can have a clean shot at the *Grandcamp*.

One of his crews begins to play out the hose from inside the warehouse. Another crew works to drop suction hoses into the murky harbor water. Onboard the ship, without warning, the deck suddenly begins to vibrate rapidly. The wood is bending and arcing, and the nails are pushing out. The cloth tarp is rising up in a panting, pregnant mound, actually lifting the timbers a few inches off the deck and then slapping back down with a thud.

Standing on the levee, a few people in the crowd begin to stare hard at the rusty hull. For a second, it looks as if the smoke and the sun are creating a mirage—the hull of the ship appears to be swelling, shrinking, swelling—some gray beast filling its cheeks with air. At 9:01 A.M., the tarpaulin is on fire, and it is shaking off the timbers.

The flaming tarp climbs into the sky like a paper bag blowing in a fog. The timbers and the hatch door buckle as if they are in a frying pan, and then they follow the tarp and blast out from the flaming hot deck.

Twenty minutes after his father left for work, Harold looks up from his schoolbook and sees the smoke swirling. He knows his father is going to be fighting this fire. When he looks up, he sees two of his best pals, Jimmy Menge and Bucky Whitley, ride up on their bikes. Like all the kids in the overcrowded Texas City school system, they are on a split shift. They have time before classes.

Let's go get a closer look.

Without a hesitation, the fire chief's son puts his books down, hops on his bike, yells to his mother that he'll be back, and follows his friends to the docks. And now, he has wedged through the crowd and found a perfect spot to view the fire from the south bank of the levee. He scans the faces on the dock, looking for his old man.

Finally, Harold spots him.

He can see his father striding onboard the ship, leading eleven other men toward the hatches.

People staring at the water lapping against the *Grandcamp* can hear a giant sizzling noise rising from the smoky, oil-stained harbor.

More people see the same thing that Bill Roach sees: The water in Galveston Bay is boiling as if slapped by some sort of monster. As if it's either the beginning or the end of the world.

The First Words from Texas City

JOHN URQUHART, Southwestern Bell Telephone Company supervisor, has struggled past the body of his chief operator, Iola Sheldon, and has frantically plugged into the toll circuit to Houston. He is shouting to no one in particular. A division traffic superintendent at the Houston telephone exchange hears Urquhart's faint, frantic plea.

They are the first words from Texas City: *"For God's sake, send the Red Cross! There's been a big explosion and thousands are injured!"*

The Scientist

APRIL 16, 9:12 A.M.

Denver, Colorado

NINE HUNDRED MILES away, a re-
served geologist at Regis College in Denver has arrived on campus and
is beginning his morning's work. He pauses to monitor his seismograph. As he watches, there is a spasm, a jolt being registered. Joseph
Downey stares at the paper charts and the lurching lines. The readings
are coming from the east. Maybe from southeast Texas, close to the
Gulf of Mexico.

The last few days there have been horrible reports from that part
of the country. A ferocious tornadic storm had formed in the Texas
Panhandle on April 9 and unleashed six different twisters along a
220-mile path that marched all the way to St. Leo, Kansas. The
largest tornado had swollen to a mammoth, horizon-swallowing 1.8
miles in width. As it gathered strength, it plucked sixty-nine lives in
Texas before speeding into Oklahoma and thundering past Shattuck,
Gage, Fargo, and Tangier. It took dead aim at Woodward, and the
city, mostly placid and unprepared, began to be destroyed at 8:42 P.M.
Winds in the Force 5 category on the Fujita Scale, ranging from 261
to 318 mph, devoured a hundred city blocks. When its work was
done, the storm born in Texas had killed 185 people and injured
thousands.

Now, with news of that immense disaster still making headlines,

Downey continues to decipher his seismograph's startling early morning reading. Downey knows there is something exceedingly powerful and unexpected gripping Texas.

It could be a massive explosion.

Maybe an atomic bomb.

The Mayor

Curtis and Edna have heard the alarm called in by Fire Chief Henry Baumgartner. The old, huge thing perched on top of City Hall can blast a signal that reverberates south to the piers and then north to where the city limits give way to empty fields.

All over Texas City, the volunteer firemen know that they have to call in to the local telephone exchange and ask the operator where the fire is. Sometimes when Edna hears the whistle she gets a friend, a woman who can make her voice sound as deep as a man's, to call the local phone operators. The woman pretends she is a volunteer fireman. That way, Edna gets to know almost before anybody else—she just really wants to know what's going on.

Just after the alarm sounds, Curtis Trahan looks out the back kitchen window of his home. The damn color blindness must be playing tricks. There is a peculiar colored smoke. Reddish orange. But that can't be right. He turns to Edna as she clears the dishes left over from breakfast and keeps an eye on the two children.

Then he realizes what he's seen: *"That looks like a chemical fire."*

Before he heads outside to his car, Curtis glances out the back window one more time. It is his turn to drive his two little boys on the fifteen-block ride south to Danforth Elementary. The shadows are still under the narrow-trunk tallow tree in Curtis's backyard. His neighbors, also peering for a few seconds out the wood-framed windows of their small, cream-colored houses in the far corner of Texas City, can see that smoke fading . . . and then just as quickly coalescing. This time it is pure crimson. Curtis never trusts his eyes. His vision, he had once been told by the government doctors before they shipped him off to the Battle of the Bulge, just wasn't what it was supposed to be.

Curtis remembers that he needs to go next door to find out if the little Colston girls, who live there, need a ride to school as well.

He thinks that being mayor in Texas City can sometimes be a good, satisfying thing.

Maybe Fire Chief Baumgartner will get a fireboat that the city won't have to sell.

Maybe kids like his office boy, Forrest Walker, won't have to go to school in split shifts because the city can't afford more buildings and teachers.

Maybe Ceary Johnson can go home to a neighborhood that has sidewalks, streetlights, and parks.

Maybe Bill Roach could finally get those whorehouses away from the homes of Florencio Jasso, Trinidad Garcia, and Freddy Vasquez.

Maybe people won't have to pass drunks and prostitutes when they go to Our Lady of the Snows Church.

Very good and very big things are going to happen in Texas City.

As Curtis puts on his tie and jacket, Edna tells him that Curtis Jr. is still too sick from the aftereffects of a bout with chicken pox to go to school. The eight-year-old is staying home that day. Their other son, six-year-old Bill, is fine and looking forward to Miss Shin's first-grade class at Danforth.

Curtis reminds Edna that he had made arrangements to drive to Houston with Basil Stewart to see if he could help Stewart get elected as the district governor of the Lions Club. A neighbor and friend, city commissioner Bill Voiles, is coming along.

Curtis likes Stewart. They are brother Lions; they can talk, even if Stewart is wedded to the oil refineries—even if Stewart is deep inside that "steel band" that Roach talks about. And Curtis also likes Voiles.

During the annexation-taxation debate, Voiles had thrown caution to the wind and voted alongside Trahan.

Just before Curtis leaves, the phone rings. Edna watches her husband with his head bowed and the phone pressed to his ear. She hears him whispering: *"I'm about to leave for work. I'll come right down there."*

After he hangs up, Curtis tells Edna, *"There's a fire on a ship down on the dock."*

He has his car keys and his briefcase. The girls from next door, Mildred and Nell Colston, are coming after all, and they all load into the Trahan family car.

It is 8:45 A.M.

Curtis makes sure the children are safely in the car. Bill Trahan peers out the car window as his father backs away from their small house and the live oak trees in the front yard begin to disappear behind them. His father steers east on Eighteenth Avenue North, aiming for Bay Street, the north-south road that parallels Galveston Bay. The anxious kids from junior high are on the softball field at Bay and Sixteenth Avenue, ready for team tryouts; Coach Galbreath is forming the players in a semicircle.

As their car travels south on Bay Street, Curtis and the children see small lines of people headed to the docks. There are dozens of teenagers—all of them up early and outdoors, emboldened by the cool breezes and the almost brisk temperature. Many of them are the students assigned to the second shift at the overcrowded public schools. They don't have to be in class until noon.

The closer his father gets to the docks, the more people seem to be converging on Third Street. There are shouts in the air. Storm-warning flags whip against their poles. There are kids on bicycles, young mothers pushing baby carriages, elderly Mexican men stabbing the ground with their canes. Cars swing wide, around the procession, and kick up the chalky-looking dirt as they aim for the dock. The narrowing streets are crowded the closer they drive to the seawall, Pier O, and the Monsanto plant. Some people reluctantly park six blocks away, almost all the way back to Texas Avenue.

On the slightly choppy bay, shrimpers looking back at Texas City can see the smoke twisting over the waterfront. At different points around the city, flocks of birds are taking wing. The shrimpers stare for a while as more and more birds take to the sky.

Mayor Curtis Trahan's young son peeks out the window again and up at the sky blue, pink, and salmon rings of smoke. The little boy thinks it is the most achingly beautiful thing he has ever seen.

Finally, his father turns onto Fourth Street and drops Bill and the two girls at Danforth Elementary. As soon as the kids leave the car, Curtis pulls away and heads toward the docks. He continues down Fourth, passing below Texas Avenue and through El Barrio. At the intersection of Fourth Street and Fourth Avenue South, traffic is jammed. People are still jockeying for parking spaces, shoehorning into little spots and then walking to the levee for a view of the fire. Curtis begins to cut his wheels to the left, thinking that Third Street might be a clearer route.

A policeman named Jim Bell has jumped off his Harley-Davidson and is directing all the traffic. Curtis rolls alongside him. He asks him a question, though he already knows the answer: *"What's the problem?"*

Officer Bell says there is a fire on a freighter. Curtis has another question: *"Are things pretty well under control?"*

Bell says that, as far as he knows, everything is under control.

Curtis looks at his watch. He suddenly remembers he is supposed to meet Basil Stewart at 9:00 A.M. so they can make their trip up to Houston.

He drives to Stewart's office at the Republic Oil refinery. Up on the second floor, Betty Benson, Stewart's secretary, tells him that Stewart is not there. He went down to the docks a few minutes ago, along with Bill Voiles. Curtis thinks about trying to hack his way through the crowds to find the men on the docks. Curtis says: *"Tell them I came in and I'll call them later to figure out when we're going to Houston."*

He decides, since he is so close by, to drive three blocks to the northwest in order to stop at the city barns, where road equipment, city trucks, and tools are stored. It's part of his usual daily round, checking in on the street and sanitation crews, seeing who is going where and what they're working on. At the long, warehouselike barn, he sees a bridge-and-street engineer named Wylie Sloman, someone who had come back from the war with a full disability after two years in a Japanese prisoner-of-war camp. Curtis stops to talk. Sloman is just about to open his car.

At 9:12 A.M., as Curtis watches Sloman put his hand on his car door, the earth begins to tremble.

There is crack that turns into a roar, as if something gigantic has snapped on top a mountain and now an avalanche of boulders is hurtling down.

The edges of the city barns are widening, spreading out, as if they are liquid and shapeless. They are shaking, melting. Jagged ten-foot panels of metal are ripping cleanly off the roof and speeding into the black smoke. Curtis instinctively lowers his head into his shoulders. He was there when the damn German buzz bomb found him in the forest. Now he tightens his body, his stomach knotted, his eyes squeezed shut. He braces for something else.

Earthquake. Atomic bomb. Judgment Day.

A concussion, the shock wave, is there in two seconds. The noise is ferocious and is hammering straight down. Curtis feels as if he is being pressed to his knees. The concussion from the explosion is pushing down on his throat, chest, and lungs. There is a vacuum, no air to breathe.

The entire city warehouse begins to collapse in on itself. A sheared-off piece of corrugated iron slices straight through thick timbers and keeps rocketing into the darkness. Curtis hears a *whoosh*ing noise of hungry fire, alive and pouncing to fill the void.

Finally, Curtis raises his head.

Sheets of flame are everywhere. There is a strange silence. And then there is a soft whirring and whistling sound. It's almost like wind moving through chimes. It is musical but coming faster and faster. Plummeting out of the rapidly rising smoke are hundreds, thousands of black, ragged things. They look like bats or buzzards. They are pieces of molten steel, and they are falling from two thousand feet high.

Curtis hears cars zooming west on Fourth Avenue, away from the docks. Chasing the cars, gaining on them, is the trembling dark ridge of the shock wave. With the flames and smoke, there is no reference point to show where the ground is, where the sky begins. It is as if the cars are driving through nothing but flames, with the black horizon racing after them. Curtis doesn't see Sloman. Then, he spots a form. The man has dived under a piece of heavy equipment.

A thought blinks in Curtis's mind: *I should have done that.*

Fighting to gain equilibrium, Curtis wrenches open the door to his new car and guns the engine, trying to find a road that can take him north.

He's got to get back home.

The City

MR. RHEEMS, the calm and tweedy chemistry teacher at Central High School, has just posed a question in his usual flat manner: *"What kind of nitrate could cause smoke to turn that color?"*

His students, bunched against the second-floor window on the southeast corner of the building, are staring at the reddish-orange swirls down past Sixth Street and somewhere over the waterfront. A student named Elaine Dupuy says out loud: *"If that wasn't a fire, that would be one of the prettiest things I've ever seen."*

Then there is a sound like a thousand empty railcars howling into a station. Mounds of jagged glass begin to hurtle, chairs are slapping down, and screams echo along the shaking hallways. Before one of the chemistry lab students crumples to the floor, he sees a white object set against the blackening sky. It looks as if it has been held frozen above the mushrooming clouds; now it is twisting, resisting; and now the object, an airplane, is plunging down into the black billows.

A FEW MINUTES earlier, in that plane, World War II Marine Corps ace Johnny Norris was banking over the pretty smoke, feeling nostalgic. He'd just stopped by the little Texas City airport to say good-bye to his coworkers. He and his wife, Betty Jo, are moving to the West Coast. He's giving up the flight game, and it seems almost impossible to believe. Norris was there in the middle of everything for a good chunk of early 1945—flying dozens of missions off the U.S. aircraft carrier *Bunker Hill,* supporting the battles on Iwo Jima, Guadalcanal, and Okinawa. Norris flew sorties over Tokyo and other heavily guarded parts of Japan, and now he is giving it all up. There isn't any money flying in and out of Texas City. The big companies have their own planes, their own pilots.

When he stops to stay good-bye, there are two customers, of all things, at the airport. They want to take a spin over the waterfront. They're offering $10 each. Norris's flying buddy R. T. Hatfield is the only pilot on duty for the day, and he's already agreed to take one of the passengers up in a Piper Cub. The other customer is still waiting. Norris looks out at the smoke. He's seen worse, and he reaches for the $10 from the other customer. He tells Betty Jo that it won't be long. As he taxis out in a second Piper Cub, the smoke is coming stronger. Norris pulls it up to fifteen hundred feet and then brings it in from the bay and inland from the docks. He waves to Hatfield, who is starting to loop out in the opposite direction, making a sweep over Rattlesnake Island.

At 9:12 A.M., the blast hits the bottoms of both planes, and they vault straight up in the air. A second later, they lurch down. The wings are ripped off both planes. They both begin to corkscrew, noses down and roaring fast for the ground. Norris's plane disappears into the black smoke and is reduced to instant rubble in a small field to the west of the Monsanto plant.

FOUR MILES AWAY, on the highway to Houston, a public bus is lifted off the ground and dropped back down. In Galveston, ten miles away, people are thrown to their knees, where they begin praying. In Freeport, twenty-five miles away, glass windows begin to pop without warning. All over the city of Houston, forty miles away, windows in homes and businesses explode and break into spraying splinters. In Port Arthur, a hundred miles away, dishes fall off display cases and windows rattle. In Palestine, Texas, one hundred fifty miles away, residents hear a boom and then feel the ground moving. Two hundred, three hundred, and four hundred miles away, up through the southwestern belt of Louisiana, geologists at Louisiana State University in Baton Rogue are reporting tremors running in a straight line connecting Lake Charles, Lafayette, Crowley, and Slidell.

In Denver, Colorado, seismographer Joseph Downey sees that jolt on his machine and wonders what the hell is going on in Texas.

As THE *GRANDCAMP* disintegrates, molten chunks of the 14-million-pound ship rage up and across the city and Galveston Bay. The tons of sisal twine in its cargo ignite into fiery coils that are rocket-launched into the sky before raining down. The multiton pieces of the *Grandcamp* begin to crash. The forty-thousand-pound deck is shot one-half mile away. Pieces of the hull are thudding into the wakes of shrimp boats in Galveston Bay, almost capsizing some of the vessels. Others are plunging onto the collapsed roofs of the shacks in El Barrio and The Bottom, immediately setting the interiors of those homes ablaze.

A raw piece of the bow crashes into the fifth floor of the Monsanto plant. The propeller shaft tunnels up from the bottom of the ship, through the No. 4 hold, travels six hundred yards, and grinds to a halt at the feet of a housewife named Olga Sawyer.

Tinged by fire, the three-thousand-pound anchor from the *Grandcamp* is airborne. Like an angry meteor, it falls out of the black smoke two miles away and gouges a ten-foot-deep crater on the grounds of the Pan American Oil refinery. A 1.5-ton drill stem is tossed from the ship for 13,000 feet and barely misses pressurized spheres of flammable butane and isobutane before it stabs the roof and then the ground floor of Amoco's No. 561 oil tank.

Warehouse O vanishes. The Southwestern Sugar and Molasses factory is incinerated, and Lutz Frieler, the company vice president, is pinned under the wreckage. His right arm is broken, his mouth cut from ear to ear and hanging down on his chest; he is thinking about the fact that massive oil pipelines run underneath the area, and every time he tries to call out, blood spurts from his chest. Along the three waterfront slips, all the oil and gasoline lines are rupturing. As each line fractures or melts, a column of liquid fifteen to twenty feet high blasts out and instantly catches fire.

More than four thousand feet away, balls of fire the size of hot-air balloons are firing out of the Sid Richardson Refining Company's oil tanks. A three-thousand-pound, thirty-foot-long oil-well pipe is airborne for two miles; then it dives and incinerates a bosomy oil storage tank owned by Amoco. Oil tanks are exploding on the property of Republic Refining. Six tanks have erupted at Stone Oil refinery. The Texas City petrochemical complex looks as if it is under attack, and now there are a total of twenty oil-storage tanks roaring with fire.

As the ship explodes, the entire basin is scooped out, and a twenty-foot-high tidal wave rises up in the north slip, as if every drop of scalding water has been knocked into one giant sheet and is crashing 150 feet onshore, in the immediate direction of Frank's Café. Plowing over the docks and toward the people stretched along the waterfront, the tidal wave is filled with diesel fuel, hydrochloric acid, ammonium nitrate, and styrene. Cresting at the top of the wave is a 150-foot-long, thirty-ton hydrochloric acid barge named *The Longhorn II*. When the wave thunders down, it slaps the block-long barge 250 feet inland, on top of a NO PARKING sign and on top of the twisted, smoking skeleton of Chief Henry Baumgartner's fire truck.

As the wave curls on top of Frank's Café, longshoreman Ceary Johnson is just stepping out the door with his coffee and leaving behind eighteen other employees and customers, including the six French sailors who have just sauntered in and are ordering sandwiches and Lucky Strikes.

The weight of the tidal wave crushes the roof; the walls crack and splinter and fly apart. The deep water hits all the people inside, knocks them down, and immediately sucks them backward at a ferocious speed, toward the fires, toward the crumbled wharf where the *Grandcamp* had once been. Johnson is facedown in a three-foot-deep sludge of acid, molasses, and oil, and all around him the tidal wave is slapping men down, grabbing them by the legs, and sucking them backward into the chemical soup of the boiling bay.

Some of the French crew are being hurled back to where they came from, back to where stoker Pierre Andre and the other French sailors are being consumed by the fire.

Thirty-two of the sailors who brought the *Grandcamp* to port, including Andre and Captain Charles de Guillebon, will never be seen again.

———

Mary Dudley turned away from the cyclone fence around Monsanto, putting her safety glasses back on and returning to her job as a lab technician. Her husband, J.L., was on the other side of the fence. J.L. had just knocked off from the night shift at the Republic Oil refinery, and he asked his neighbors—Harold Fletcher and his dad, Emory Fletcher—to join him on a trip to see the fire and to check on Mary. They all met at the Monsanto fence.

Now a wall of glass is coating Mary as the explosion erupts. She is blown backward, and her face and arms look like wallpaper that has been raked. Her eyes are protected by her safety glasses. Her husband sees a red flash but hears nothing. Then J.L. is flying, airborne, tossed west down the dock road. He tries to stand and sees that his shirt has been blown completely off but his jacket is still on. He is wearing just the waistband and one leg of his pants. And then he notices, rising from the pools of mud and fire, dozens of black people.

"Where did all the black folks come from?"

His friend and neighbor Harold Fletcher is picked off the ground and thrown into the eight-foot-high cyclone fence. The fence instantly rolls shut around him, squeezing him into the metal. He hears nothing, and he cannot breathe. Then there is a loud whistling coming from the Monsanto plant. Pieces of the building and body parts begin to thud down around them. Fully intact shirts and pants flutter to the ground.

Escaping and crawling on his hands and knees, Fletcher looks for his father. He looks toward the Texas City Terminal Railway tracks and sees a blown-out skeleton of a boxcar.

A naked man is holding on to the smoldering frame of the car. When the man smiles, Harold sees some gold teeth and realizes it is his father. His side had something blown through it. Harold scoops his father up and turns to walk away.

There is a man sitting on the sidewalk with his gaze fixed on something in the distance. Harold realizes the top of the man's head is gone.

Alongside Monsanto, a woman is staring at them. She is naked. Harold looks down and realizes that he is wearing only the waistband of his pants and that somehow his pants pockets have been seared into his skin. Coarse black liquid is embedded in his body; it feels as if it has been blasted permanently into his skin. Harold notices something else; there is a silver-dollar-sized hole in his leg. Harold carries his father through some fiery ditches on the edges of Monsanto, then stumbles and can't move anymore. He lays his father down, waits for help, and goes into shock. His father is loaded onto a truck and taken to the hospital in Galveston, where he is put in the "colored ward"—along with all the other people who are so badly burned and coated with the black molasses–oil mixture that nurses and doctors assume they are Negroes.

On the scorched grounds of Monsanto, where 451 out of the 658 employees have reported to work, each of the plant's major buildings is on fire, crumbling or flattened into a glowing rubble. There is a sharp, acrid smell of acids and petrochemicals. Churning spouts of benzol, ethyl benzene, and propane are shooting out of ruptured pipes.

Caught inside Monsanto, in the chaos of falling girders, melted steel supports, and collapsed walls, engineer Fred Grissom feels like a rag doll bouncing up and down. Then there is the utter, strange silence. It is suddenly so quiet that he can hear the drips of blood falling from his face, from where his right eye used to be, and hitting the ground. For Grissom, it is like looking through a dense red fog. He puts one foot forward and realizes that he is touching a body. He falls to his knees and presses his face six inches away from the fallen figure. Through the blood in his left eye, he examines the hair color and the shirt. Though he can't recognize the mutilated figure, he determines it must be one of his best friends, Tom Womack.

Grissom begins to walk out, his arms outstretched to feel his way. He hears a woman screaming for help. He follows the voice, reaches down, and scoops up the immobile woman. As the nearly blinded man carries her, she guides him through the fires. In a few minutes, Grissom is in John Sealy Hospital in Galveston, and doctors are leaning in to him, whispering to him after his morphine shot, telling him that they are going to have to remove what is left of his right eye.

Calvin Martin, who works in the Styrene Distillation Units 5 and 7 with his friend Milton Skillian, is just taking off his tool belt and opening the door to the shack where employees are allowed to take a cigarette break. He is following Skillian inside. As he pushes the door, he hears a sound like a brutal wind, like a blue norther. Martin looks up and sees his friend Skillian flying out the back of the smoke shack, thirty feet in the air, in a sitting position. Martin blacks out until some-time later he comes awake, listening to a voice he doesn't know saying, "Here's an arm."

On the second floor of the plant, in the polystyrene unit, John Davis wishes the noise would just end. It goes on and on, and it is a mixture of falling buildings, cascading black rain, and screams. He is unclear exactly how it is that he is going to die.

I must be drowning.

Machinist's apprentice Joe Dillon is in the tool shop, putting his trembling fingers to his head. It feels like soft fruit. He has been in the Battle of the Bulge, the Ardennes Forest, across the Cologne Plains in an army tank. He had never been injured, until now. He lies still, wait-ing to die.

Not much use in trying to get out.

A few seconds later, he presses his fingers to his head again and this time it feels solid. His leg and pelvis shattered, he begins to pry himself out from under bricks and machine parts. He crawls into the only safe place he can find, an exposed chemical pipe. He will need thirty-seven blood transfusions and will require a shot of penicillin every two hours for the next ninety days.

Monsanto storeroom clerk Madelaine Rockefeller is being slammed by cascading bricks. She drops to her knees. Assuming she is dying, she closes her eyes. There is an enormous wetness passing over her, a mix-ture of oil and chemicals. She opens her eyes and sees a hundred-foot square of fire in front of her. The two twenty-story petroleum distilla-tion towers are barely standing, both at odd angles. Benzol storage tanks are burning. The styrene tanks look as if they have been stepped on. The plant's main storeroom has only two walls standing. Mon-santo's dockside warehouses have disappeared. The boiler room, where Forrest Walker's father works, is a piled-up mass of shattered brick and tangled supports.

Rockefeller crawls outside the Monsanto plant, where a dozen horri-bly burned men are wading through twin pools of chemical-laced water coated with dancing fire. Her supervisor, plant superintendent

H. K. "Griz" Eckert, is unconscious and under a mound of debris. He has several ounces of glass splinters imbedded in his head.

The dead bodies of seventy-two workers are scattered over Monsanto's forty acres. For the time being, eighty-two other full-time Monsanto employees have vanished somewhere in Texas City, in Galveston Bay—somewhere inside the crumbled masonry and steel girders that are bent and twisted like the flimsy wires of a birdcage.

The Mayor

IN FRONT of Curtis Trahan's car, thirty women and children are standing in the middle of Texas Avenue, screaming. He jumps out, waving his hands. The air now smells of sulfur and charred flesh. There is a car, its windows melted like cellophane but with no other signs of damage. The man and woman inside are decapitated. Curtis stands in front of the crowd in the street:

Don't panic. If you've got a car, just go ahead and get out of town. I have no damn idea what's happening.

Curtis steers down Texas Avenue, and there are crowds spilling out everywhere, running away from the docks, from Monsanto, from El Barrio and The Bottom. Shreds of cotton are falling from the sky, sometimes in dense enough patches that it looks as if it is a blinding snowstorm. The line of fleeing people, Curtis guesses, is a mile long.

El Barrio is virtually gone and Trinidad Garcia, the janitor for The Company and the man who once rode with Pancho Villa, is searching for his family. Some of them have simply begun running out Bay Street, aiming for the Texas City dike; and now they are sprinting along the little finger of land into Galveston Bay, hoping that the explosion doesn't corner them at the end of the dike.

With the pushing black smoke at his back, Curtis turns right on Sixth Street and heads north. There are three people stepping out of a collapsed two-story building. Their faces are coated white, and they are coughing, sputtering, and trying to flick asbestos from their hair, faces, nostrils, and mouths.

The enormous plate-glass window at Texas National Bank has just slammed Dr. William Anderson over the head; the bloodied physician is now on his knees. Two shoes are sitting aligned perfectly alongside each other in the middle of Sixth Street, as if someone has been blown right out of them. Elementary school teacher Lubelle Belcher is wandering the street; glass shards are imbedded in her eyes and blood is pouring from a wound in her head. At Nuchol's Grocery, a clerk named Louise Holland has just crawled out from under the broken windows and roof debris and is throwing her crushed eyeglasses on the ground. The front window of Butler & Grimes Five & Dime has been neatly sliced from its frame and is hovering over the head of a customer who had just turned the handle at the front door. Men are in front of the liquor store, retrieving unbroken bottles of bourbon and taking big slugs.

There is a toxic smell everywhere from the petrochemicals, the toluene, the styrene leaking out of burning towers and railcars. Dozens of cars are on fire, stacked on top of each other like firewood. Along the south end of Bay Street, close to the bay, the sky is turning the color of Mercurochrome. There is a curtain of fire and it has black, lacy edges around its red, yellow, and white center. A young child is sprinting down the street, reaching to hold the hand of a stranger, a woman, running alongside . . . and the woman's legs are starting to wobble . . . and the child looks up to see that the woman has been running even though she has been decapitated.

Pat Manis, twenty-two, throws her laundry down at Hipp's Laundromat and drives straight to the Union Carbide plant to find out if her husband, Johnny, is alive. Close to the chemical plant gates, there is a man covered solid with oil and she stares, horrified, as he begins to scream:

"They're all dead. They're all dead. They're all dead."

Manis slows for a second and asks the man who he is. He doesn't seem to hear her and he continues to scream:

"They're all dead. They're all dead. They're all dead."

Fifteen-year-old A. C. Calhoun, who had been working on his bike at Shannon's Bike Shop, is ducking because chunks of metal the size of automobiles are airborne over Texas Avenue and going uptown. With a sixteen-inch gash on his leg, losing blood fast, he begins limping past Texas City National Bank and sees the vault completely exposed now that the walls are gone. Watches and diamond rings are scattered over the sidewalk.

A woman with an infant runs up to the teenager: *Please help me.*

The teenager looks down and sees that the jugular vein on the baby boy has been severed: *"Lady, I can't help him."*

OUT ON SIXTH Street, Curtis Trahan is thinking about his boy. *Where the hell is Billy? What happened at the elementary school?*

On the sidewalk, there are streams of blood around the front doors of the Danforth Clinic. The chemical smell is dizzying, overpowering. People are pressing wet rags to their faces, trying to remember what soldiers in World War I had done during the mustard gas attacks. As he passes his little insurance agency office, Curtis can see all the glass blown out. Frantically, he scans for familiar faces.

There is Forrest Walker, the teenager who works for him.

What the hell is he doing here?

Walker is waving. He has Billy Trahan by the hand. He had spotted the six-year-old crying and wandering with a group of other first-graders in the wreckage fanning out from the elementary school on Fourth Street. Walker had grabbed the delirious boy by the hand and walked him up the street to Curtis's insurance office.

Walker looks at Curtis. *"Mr. Trahan, I don't know where my father is."*

Curtis tells Walker to go, right away, and find his family. Curtis loads Billy into his car and continues home. The usual five-minute drive has turned into a nightmare crawl. Drivers are careening to avoid bodies that have slumped in the middle of the street. There is the constant bleating from car horns and sirens. A butcher from Cook's Grocery is huffing down the street, still wearing his apron. Julia Turner, a driver for Monsanto, is also running away from the docks. All of her clothes have been blown off. A piece of plastic and a safety pin from her stockings have been seared into her skin. Her naked body is coated with something tarry and thick, and like many people fleeing the waterfront, she looks like a black person. She begins screaming, realizing she is alive: *"Thank you, Jesus! Thank you, Jesus! Thank you, Jesus!"*

As he drives, Curtis spots six-year-old Joan Voiles, the daughter of his neighbor, City Commissioner Bill Voiles, the man who had decided to vote with Trahan during the annexation-taxation debate. He yells for the girl to get in the car.

Now.

He has a bad feeling.

When Curtis had gone to find Basil Stewart and Bill Voiles earlier that morning, the secretary at the Republic Oil refinery had told him that the two men had gone down to the docks . . . to the fire. God only knows what the hell happened to the little girl's father.

With Joan in the car, Curtis speeds away. In his rearview mirror, the entire sky is filled with a black mushroom cloud. It looks like Nagasaki or Hiroshima. Curtis is at the northern edge of downtown, turning to the northwest and racing to his house.

In front of Michael's Jewelry Store, it looks as if the entire building has been pushed back from the sidewalk while, at the same time, its entire contents have been hurled to the street. A man is bent over there, filling his hat with spilled diamond rings.

EDNA IS IN the bedroom, pulling the sheets tight, fluffing the pillows, arranging the bedspread. It is cool, brisk; there was a brief rain overnight, and the air seems scrubbed.

She yells, over her shoulder, for eight-year-old Curtis Jr. to go in the kitchen and turn down the gas flames on the stove. Then the explosion comes, jolting the house, shaking the ground. She freezes in fear.

What have I done? I just told my boy to go to the stove.

She runs to the kitchen, sees wide-eyed Curtis Jr. standing there, then sees that the stove is still intact.

It's the water heater.

She happens to take a quick look out the window, and she sees the utter black sky. The back door is blown wide open. Edna grabs Curtis and checks her house, going room to room. She steps outside and circles the exterior. The smoke is climbing. Her neighbors are out in the street, pointing south at the mushroom cloud. Finally, she hears car wheels screeching, a car horn honking, from around the corner.

It's Curtis. He has Billy in the car. He is flashing the lights, and he is yelling: *"Get in the house. Get in the house. Do not leave this house. This town has just gone berserk."*

Edna reaches for her little boy. She pulls him to her chest and rocks him back and forth. Billy is still engulfed in tears and mumbling something about the ceiling in his school tumbling down, and about all the windows being broken.

Edna doesn't want to believe that any of this is happening. She thinks, for a few seconds, that Curtis is the one acting crazy. Curtis isn't like this. Out on the front lawn, people are racing by, some clutching babies. Commissioner Bill Voiles's wife has come over from her white bungalow, and she is carefully asking Edna if she knows what has happened to her husband. He was supposed to go with Curtis to Houston.

Edna's relatives run to the house. Her sister Inez is bleeding. When Inez heard the explosion, she instinctively reached into her infant son's crib and put her hand, palm up, over the baby's head. The walls trembled, the glass came flying, and a shard of it pierced her palm all the way through. When she lifted her bloody hand, she saw that only the needle-sharp tip of the broken glass had hit the baby, who now has a wound in the shape of a star on his forehead.

Edna goes inside the house. Billy, the first-grader, seems calm. Together, they sit in an old chair by the window that faces out on the corner where Ninth Avenue meets Nineteenth Street. Suddenly, tears begin sliding down Billy's cheeks again. She follows his line of vision. Outside there are even more people running aimlessly through the neighborhood. Most of them are crying. Some are hobbling. Several have bloodstains soaking their shirts or matting their hair.

Edna is horrified, and she thinks it is something from the Bible:

"It's like a scene from Exodus."

Pressing her wailing son to her body, she reaches up, draws the blinds, and wonders what has happened to all the other children at Danforth Elementary.

PRINCIPAL RAY SPENCER is shaking his head, adjusting his tie, and trying to catch a breath after chasing ten-year-old Raymond Dupuy and all those other kids—he has forced them back to school after finding them sneaking down the road to the dock. As he herds them inside Danforth, the first explosion pitches him to the floor. Principal Spencer's arm is almost severed and blood begins to pour into his shirt-sleeve and jacket. Children are rolling, jerking, and the walls begin to cave onto them.

Rheba White is in her first-grade reading circle, just beginning *Three Billy Goats Gruff*. The east bank of windows lacerates her class-mates. A glass shard punctures Rheba's hand. Outside her classroom, the hallways are filled with clouds of plaster dust. One boy, who uses steel braces because of his polio, is being hoisted onto the shoulders of a schoolmate. In the adjacent classroom, substitute teacher Eloise Cameron is blown to the floor. She reaches for the doorknob, and it has been severed. Seven- and eight-year-olds are crowded around her. They are mute for a few seconds, and now they are screaming: *"I want out! I want out!"*

Beryl Wages, the fourth-grade teacher across the hall, yells for the students and Cameron to put a desk under the transom so the kids can crawl out the little space. One by one, they tumble into the destroyed hallway and begin to crawl through the fallen lockers, blood-coated bricks, and glass. At the stairwell there is a frantic bottleneck.

Dozens of children are packing into the narrow space, and dozens more are desperately pushing them from behind. The lower level of the stairs has collapsed, and it is blocked. Principal Spencer, with one good arm, is trying to peel away a path for the kids to follow. Beryl Wages sees two little girls being trampled in a bend of the stairwell. Wages's face is gouged, and her ears are ringing with the constant, high-pitched screams.

Children clamber up a four-foot-high mound of rubble in the fallen stairwell and then slide down to a spot where they can jump another four feet to the granite floor. As the children stumble into the playground, they are converging with students coming from the adjacent high school. There is a furious stampede. Several kids begin to climb the school fence. Some of the first-graders are silent. They sit down, folding their dresses and shirttails underneath them. Then they place their hands over their ears and begin praying:

"Good Lord Jesus, deliver us from evil."

Outside, first-grade teacher Eva Howard is looking into the dazed face of Hazel Wafford. The little girl isn't supposed to be in school until the afternoon, but she has fought her way into Danforth after the explosion. The seven-year-old's face is covered with oil. She is out of breath from running all the way from the docks.

"My daddy was on a boat, working. He was thrown in the air. I saw him."

CURTIS LOOKS AT his wife and sons and says he has to go. He demands that they stay put. His voice is booming, and he looks as if he has the fear of God imprinted on his face. Edna thinks she knows her husband, and she has never seen him like this before. She doesn't want him to leave. As he steps over the broken glass and goes outside, he repeats himself: *"Do not leave this house. Do not go outdoors."*

Now he is back in his car, tires spinning, retracing the same route back to downtown. As he comes down Sixth Street, each block bringing him closer to the southern end of town, there are bodies lying along the sidewalk. A determined-faced man in a suit, his head held high, is walking in front of the Showboat Drugstore with his arms folded across his chest. His stomach is gouged, and he is trying to keep his intestines from falling out. Another man, of indeterminate age, is running from the smoke. He is naked except for a belt around his waist. His body is the color of iodine, and he is covered with oil and blood. A woman with crazed eyes and blood blossoming in wide patches on the front of her sundress is walking barefoot through the broken glass, holding a puppy and begging for someone to wake the dog up.

Ahead, at the far end of Sixth Street, the smoke is pressing down over El Barrio and The Bottom. Curtis knows that those neighborhoods are the closest to the north slip and that there might not be anything left. As the winds carve small, murky windows through the black haze, he can see row after row of swaybacked homes with roofs and walls that have been sucked down and in, as if their contents have been vacuumed out.

It is 9:40, and Curtis is finally in front of City Hall.

The building looks like something from the grainy World War II newsreels that Texas City children had been viewing at the Showboat Theater. Windows are cracked; doors are at odd angles; desks, file cabinets, and chairs are knocked to the floor. Curtis can see rows of people—some being carried, some being pulled in wheelbarrows, some hanging on to the hoods of cars—approaching City Hall from the south. Even with the lower end of City Hall looking as if it had been hit by a missile, there really is no place else to go. Wounded people are crawling on the sidewalk. The grass is matted by thick splotches of blood. A woman is slumped under a tree; she is pressing her right hand to the left side of her face, fighting a losing battle to keep her eye in its socket.

As he hurries inside City Hall, Curtis passes three people coming out. A young woman and funeral home operator Fred Linton have their arms wrapped around a man. They are walking him out, and his feet drag along the ground. He seems to be about five feet seven inches and 130 pounds. He is conscious but immobile and silent. His skin is covered with that congealing oil, tarlike molasses, and chemical residue. Strangely, his eyes are completely open; he is staring at something only he can see.

Curtis runs his hands over the man's face, gently wiping aside the blood and oil from the man's lips. The tender mercy doesn't bring Father Bill Roach to attention.

Curtis stares at his dazed ally. He sees no open wounds; nothing seems broken. With the walls of heat, the piercing sirens, and the babble rising up from Texas Avenue, Curtis suddenly wishes he had a towel or a brush with which to clean his friend.

He reaches for his comb and begins to run it through Bill's hair.

The People

APRIL 16, 1947, 9:12 A.M.

Julio Luna, the Longshoreman

JULIO LUNA follows the line of sweat-stained longshoremen past Frank's. Three of the unshaved French sailors are already inside, clamoring for some last-minute packs of cigarettes. Someone's offering to drive Luna and four members of his gang to the International Longshoreman's Union Hall. Luna says he's ready. Until the fire is put out, there isn't going to be any work anyway.

The work whistles have been howling nonstop, and crews are sliding down from the other two rusted ships in port, the *Wilson B. Keene* and the *High Flyer*. A sailor from the *High Flyer* crew has strolled over with a camera and is snapping photographs of Chief Henry Baumgartner's crew desperately unreeling the hose and playing it on the deck of the *Grandcamp*.

Luna's car rocks out on the dock road, and, at 9:12 A.M., the rear of the vehicle is forcibly yanked up toward the clouds. The explosion is gripping the car, holding it, crunching it, and in front of it burning wheels and glowing metal spears are cascading and attacking the ground. Instantly, there are cliffs of flame on all sides of the car. The car slams back down and lurches through the fire, going north. Luna turns his head and watches three-story-high fuel tanks igniting. The car spins out to the loop around the city. Finally, the car caroms to a halt. Luna and the other longshoremen are breathless.

It is hell on earth.

They decide to go back to the waterfront to rescue anybody who is left alive.

Mike Mikeska and Walter Sandberg, The Company Men

Mikeska is livid.

He has already told Sandberg to go check on the goddamn tugboats from Galveston. Sandberg went back to The Company offices dockside, and there still isn't a single tug on the horizon.

Mikeska has no way of knowing it, but the crew of the *Albatross*, a tugboat owned by G&H Harbor Tugs, didn't shove off until just after 9:00 A.M. for the fifty-minute sail to Texas City.

In The Company office, Sandberg is still calming the edgy owners of the *High Flyer*. Their freighter is docked in the main slip, and it already has nine hundred tons of ammonium nitrate in its holds.

Sandberg cradles the phone and explains that there is really no danger, that everything is under control. Then the explosion blasts him to the office floor, and files, papers, desks, and chairs come flying at him.

Mikeska's secretary Grant Wheaton, who just handed Sandberg the phone, looks as if he is dead, lying on the hardwood floor. Sandberg calls out to Wheaton, and Wheaton finally says he is okay. The men crawl, and Sandberg realizes that Wheaton is not okay after all. He is going to lose his right eye. Sandberg puts Wheaton on the ground and tries to start a car near what is left of The Company headquarters. It won't start. He tries other cars. None of them turn over. A pickup thuds by with four bodies stacked in the back. Sandberg flags it down and loads Wheaton onto the truck.

Sandberg is beginning to realize that he might now be in charge of The Company.

There is no way in hell that Mikeska could have survived.

The tugboats Mikeska ordered dispatched from Galveston are near Bolivar Peninsula and the tug crews brace themselves against the booming echo from Texas City. The tug captain reaches for his two-way radio and is told that the *Grandcamp* has just exploded. He decides to steer closer to Texas City anyway. As he moves through the cut between Galveston Island and the peninsula, he can see the fires and the smoke blowing in his direction. Moving into the ship channel, he sees people flailing in the water. The *Albatross* steers through dead bodies and body parts bobbing in the shallow bay. J. D. Babin, an oiler on the *Albatross,* thinks the scene is spectacular and terrible. He's thinking the same thing as Sandberg: There is no way anyone who was close to the *Grandcamp* could have walked away from it.

Mikeska, like Bill Roach, was staring into the full brunt of the smoke and the heat when the *Grandcamp* exploded.

Mikeska—the man who ran The Company, who inherited it from the colonel, who never wanted Texas City to annex the corporations on the waterfront—will never be seen again.

On Ninth Avenue Mikeska's daughter Beth has forgotten all about the coffee she's brewing. She runs into the street, into the stream of bloody, fleeing people. There is that formless stream of noise—screaming, explosions, sirens. She is hoping for any news about her father.

Florencio Jasso, the Merchant Seaman

Jasso is always amazed at the lack of security around the docks, at the way that almost anybody who wants to get close to the waterfront can find a way to do it. As the seconds click to 9:12 A.M., he has seen several more friends, men who worked at The Company's warehouses, and he joins them on the side of the dock road facing Warehouse O.

World War II veteran Jasso is standing seventy feet away when the hulking oceangoing *Grandcamp* instantly becomes a blinding mountain of white fire.

He feels himself being lifted off the ground, hurtling backward, his ears completely filled with a shuddering buzz, and then he sees a high-voltage line snapping up and down in front of him and the towering grain elevator shaking. The friend who has driven him to the waterfront has disappeared. Jasso is now on his back in a shifting gully, and then he forces himself to his feet and tries to churn his legs. He nervously touches his head to see if it is intact. His shirt is gone; his pants are shredded.

Jasso runs and runs and finally finds what he thinks is the little Mexican American area of railroad shacks built by The Company. His house is cleaved in two, and flames are licking around it. He can't find his mother; then, finally, he sees a woman about a hundred feet away, tumbling through the smoke.

She is falling, or something seems to be hitting her knees. Jasso runs to her side, props his mother up, and then carries her to Fourth Avenue, in the heart of El Barrio. There are trucks there already, taking bodies to the Danforth and Beeler-Manske Clinics and then to the Galveston hospitals.

His mother will have her leg amputated.

He is deaf.

His father, who helps to keep The Company railroad cars oiled, has climbed onboard one of The Company engines and hung on as the crew slammed the whole thing in reverse and tried to chug away from the docks.

It will be three days before Jasso's father learns that his son and his wife are alive.

Forrest Walker Jr., the High School Senior

The high school quivers from the first explosion. Three seconds later another wicked, bone-jarring explosion rips the air. Forrest is trying to crawl under his desk. Meyers, his classmate, is already down there, screaming about the welling pressure in his head: *Open your mouth. Open your mouth.*

There is a piercing pain in their ears, as if something is drilling right into their skulls. Hundreds, maybe thousands, of people around Texas City are feeling the exact same pain in those exact same seconds.

Escaping from that high school typing class, there is little overt panic. People jog briskly, in silence, past the outside door that has been blown off its hinges. Outside, things are worse. There is welling chaos as the high-schoolers merge with the screaming elementary school children. A teacher tells Forrest to run to Sixth Street and get the pharmacist and any supplies. Out on the main road through town, Forrest can hear people already talking about how Texas City needs to be utterly abandoned.

After he rescues Billy Trahan and delivers him to Mayor Curtis Trahan, Forrest finally finds his mother. Their house is two miles from the dock, but it has suffered. Nails are popping out of wallboards; an entire window frame is punched out onto a bed; the refrigerator is gouged by flying debris.

Forrest and his mother load into the 1940 Pontiac and speed north, following the road out of Texas City to Houston. Forrest had seen those wartime newsreels in the Showboat, where he is one of the movie ushers. It looks like footage of European refugees wandering through rubble. He and his mother pick up an elderly couple, in their seventies, and they press on. The old man in their car says that he had been in bed with pneumonia but that he had begun staggering away from Texas City as soon as his neighbors told him to flee.

Twelve miles up the highway, they finally stop in the small bayside town of Kemah. Dozens of other cars from Texas City, part of a long, escaping caravan, are already there. Strangers take Forrest and his mother in. The elderly couple moves on.

All afternoon and into the night, Forrest listens to the radio reports for the list of the known dead, for news about his father.

John Hill, the Chemical Engineer

Just as his eyes close, the bed begins to rattle violently.

Still woozy, he steadies himself and goes to the hall heading to the kitchen. He passes the back bedroom and looks out the rear window. A ridge of quavering black air is hurtling from the southern end of Texas City. It is ten feet off the ground and stretching for miles. The rumbling black shock wave is racing straight for Hill's house.

There are bodies cartwheeling in the sky.

His house rocks, creaks, and ultimately holds steady. As the black mass passes, his wife is trembling against him. When there is silence, when he knows that his house is sound, Hill yells that he is going downtown, to City Hall, to find out what the hell is going on. As he speeds down Sixth Street, hundreds of people are walking toward the Danforth Clinic. He parks near City Hall, runs up the steps, and finds Curtis Trahan.

Curtis has just put Father Bill Roach on the medical evacuation bus to Galveston.

Curtis gives Hill a briefing: The three hospitals in Galveston have been notified. Somebody went out to the airport to put out any fires that might prevent a plane from landing. There is another ship, the *High Flyer*, that people are worried about. He has no idea what the hell has happened to Fire Chief Henry Baumgartner, but the early reports aren't very good. El Barrio and The Bottom are uninhabitable, almost leveled. The sprawling Monsanto plant, spread across forty acres, simply doesn't exist anymore. Most of the warehouses along the waterfront have disintegrated.

Hill studies Trahan. They are friends from playing bridge, from hanging out at the Lions Club, but Hill has never liked Trahan's politics.

"Curtis, you have to sit down and get this damn thing organized."

Trahan is beginning to fathom what lies ahead. *"Well, all right, but in the meantime, just do whatever the hell you need to do."*

Curtis adds that they will meet later in the day to sit down and map out a specific plan to corral everything.

Right now there are immediate duties: Standing in the middle of Sixth Street and barking out orders to people who are walking down the middle of the business district. Carrying the wounded or dead out of the smoke. Herding any of the bloody children from Danforth Elementary. Beating back looters who want to invade the Texas National Bank. Screaming out directions so the Houston and Galveston fire departments know where the hell to go. Shoving the striking Southwestern Bell Telephone operators back to work.

As the two men frantically huddle in front of City Hall, wild-eyed strangers are running up and shouting: *We need help with some flat tires.*

Curtis stares back. Someone else is yelling out the window of a car that looks like it's been hit by a flamethrower: *Where do we bring the dead bodies?*

As Curtis steps away, Hill runs to find an open phone. He finally manages to patch through to Marge. He tells her to escape, with their son, and go all the way to Houston.

He'll be staying downtown for a long time.

Curtis has told him that, for all intents and purposes, he wants the twenty-eight-year-old to be the deputy mayor of Texas City.

Elizabeth Dalehite, the Sea Captain's Wife

As she prays, she sees Henry walking into a wall of fire.

At the same instant, the head on the statue of Blessed Mary she's holding is severed straight from its body. Now there is a deep cut on Elizabeth's hand, the one that is clutching the headless statue.

The force of the blast sucks Elizabeth straight out of the window of the car and toward the row of wooden shacks where Flo Jasso and his parents live. As she tries to regain her feet, there is someone frantically grabbing her around the neck. A Mexican American woman has stumbled over from the shacks and is grabbing anything, anyone. Elizabeth shakes the woman off and tumbles back to her car. The hood is up; the doors are still closed, and it has been knocked into a ditch. She climbs inside. Everything has been blasted out of the car. Books. Papers. Cigarettes. Even the car's ashtray. The interior and exterior are covered with singed, smoking wires and spools that have dropped out of the sky. Suddenly there are men surrounding her, shrieking at her: *"Can you drive?"*

Elizabeth sees the men cradling bodies in their arms.

"Yes."

She is at the wheel, the car is still jammed in the ditch, and the men are flinging the back doors open and shoving wounded, moaning people inside. Someone else is cutting the knot of wires, trying to free her car. Her mouth doesn't feel right. Her teeth have been smashed down into her gums by something. Her head is bleeding. Through the oil on her body there are little sparkles from glass fragments in her skin. Gulping air, her chest hurts, as if someone large had brutally stepped on it.

She hears someone screaming: *"Drive to the hospital!"*

Slipping into nervous shock, Elizabeth manages to steer past burning bodies and fiery pieces of metal. She finds the road to the Danforth Clinic. The bodies are unloaded. A man at the clinic takes one look at her and rams some gauze into her hands. Elizabeth stares down at it and then shoves it back at the man.

She runs outside, dodging the cars and the bodies, trying to concentrate. Now blood is coming out of a gouge in her finger. She'll go see the Bynums, another family, like the Dalehites, that helps run the ships and boats out of Galveston Bay. They're friends. They'll know what to do. They'll know where to find Henry. Maybe the phone will be working at their house.

Elizabeth careens onto their front lawn. Water is gushing out of the house, pumping out of the doors and cracks of the splintered building. Pipes have been snapped, and the Bynums are wading through the deluge inside. They grab Elizabeth, pull her indoors, and hand her the phone.

Who should I call? Should I call Henry Jr.?

Her eighteen-year-old son, a student at the University of Texas in Austin, had worked as a deckhand for his father, rescuing the boats when the hurricanes blew in. His father was proud when his namesake decided that he was going to go to law school. Henry had been going to sea since he was ten. He had seen his son grow up on the ships. Now he hoped that his son was going somewhere else. There was even talk that Henry Jr. might go up east, maybe to Columbia or Harvard . . . maybe even to England to study law . . . somewhere as far away as you could get from the oyster reefs alongside Offat's Bayou and the sandbars off Galveston Island.

Elizabeth quickly decides to call the seasoned crewmen who usually go to sea with her husband. Standing in the house, water swirling everywhere, she is convincing the local operator to please patch her through to the office of the Dalehite Boat Line. The line finally goes through, and now she is begging: *"Please . . . send me help. Somebody come and help me hunt for Henry."*

She hangs up, and she can't just stand there. She runs back outside, her clothes soaked with oil, water, and mud. She wants to go back to the dock. Elizabeth walks south, determined and ignoring the blinding smoke. The police are scrambling along Fourth Avenue, waving people toward open paths through the shrapnel and the smoldering, two-story-high hunks of the *Grandcamp*'s hull.

There is no sense that it is inching toward midday, that it is still an unusually clear day along the rest of the Gulf Coast. The arc of black smoke is so encompassing that it looks, to most people, as if it is the dead of night.

There is a man coming toward her, someone who seems to recognize her. He is bleeding and reaching for Elizabeth: *"Mrs. Dalehite, get out of here as quick as you can."*

She finds the police line and pleads to get through. There are people still running from the dock. Elizabeth watches women with bundles that could be children. They are hunched over, low to the ground, their bodies tightened to the possibility that another explosion may be coming. The policeman won't let her through. Elizabeth begins to walk, by herself, to the outskirts of the city, out toward the curving intersection where the road forks and splits to either Galveston or Houston.

Henry had told her that one time he had to tow a gasoline barge from Houston to Corpus Christi and that they couldn't talk to anybody about it—it was a military barge, something for the government that needed to get down the Texas coast as soon as possible and as quietly as possible. He stood at the ship's wheel for forty straight hours as it dotted its way through Galveston Bay and the Gulf of Mexico, stopping occasionally to pick up new cargo and getting new instructions. At every stop, even at 1:30 and 2:00 in the morning, Elizabeth was there to bring him coffee and food.

"Time meant nothing to me when I was with Henry. I devoted all my time to my husband. Time did not mean anything to me."

Kathryne Stewart, the Mother

As she runs the comb through her bangs, there is a terrifying thunderclap. Then another. Then one more.

My eardrums are collapsing.

She sprints outside, and there is an insistent, horrible drone coming from the waterfront. The southern sky looks like the photographs of the atomic bombs in Japan. Dense waves of black smoke are mushrooming, multiplying. She runs to the car at the curb. Stu has been knocked to the floor of the car, but he seems all right. Across the street, the home belonging to a Monsanto worker, his young wife, and their three-month-old baby looks as if it has been torpedoed. The front walls are gone. The living room and other parts of the home are exposed to the street. Kathryne sees the young mother. Alive, she is futilely trying to hide in a corner of their exposed bedroom, clutching her infant to her chest.

Kathryne wants to scream, and, of course, nothing emerges.

Marilyn. We must go see about Marilyn.

With her five-year-old in the car, she makes her way through the carnage on Sixth Street, following the same path that John Hill, Curtis Trahan, Elizabeth Dalehite, and others are following. She turns on 4th Avenue and is watching the panicked children fighting their way off the playground at Danforth Elementary. She cannot hear herself think. There is a deafening rumble from the dock as the flames continue to roll and tumble. Not a single window on the small elementary school is intact. She fights her way against the tide of children, finds Marilyn's room; it is deserted.

The floor looks like a crimson and red quilt of blood, first-grade storybooks, spelling tablets, glass.

She grabs Stu's hand.

"Whatever happens, don't let go of my hand."

Back outside there are frenzied collisions with children, parents, teachers. She scans the faces as the first-graders and second-graders try to shove by her. She sees Marilyn's teacher. The woman says she has no idea where Marilyn is—but that the last time she saw her, the child was unhurt.

Kathryne leans against the wall, thinking that she is about to collapse, and she rubs the back of her hand over the tears coating her cheeks. She sees Marilyn's teacher going into what is left of the school nurse's office, slamming the door closed behind her. As the door shuts, Kathryne can see a hysterical, blood-soaked six-year-old girl inside.

Not knowing where else to go, Kathryne guides Stu through the mob. They find their car and retrace their route home.

Fifteen minutes later, a neighbor is at the door with Marilyn—she had been picked up wandering, like Curtis Trahan's son Billy, past the corpses and body parts splayed across the streets closest to the waterfront.

Marilyn is quiet. Kathryne holds her daughter and decides not to press her about what exactly the child has seen and done.

Her relatives begin to arrive.

Is there any news about Basil?

Only now does she begin to completely absorb the fact that there has been absolutely no word from her husband. No news about him from anyone who might have been with him downtown. Nothing from their friend Curtis Trahan, the man whom Basil was supposed to meet this morning.

Kathryne turns on the radio. She hears announcers reading a tentative list of people assumed dead. She listens. Coworkers of Basil's are mentioned. Some friends. Basil's name is not mentioned.

"We have just read you a tentative list of the identified dead in the Texas City Disaster. . . . We urge you to consider that this list of names is not official and several identifications have proved to be erroneous. If the name of a member of your family was read, please check immediately with the Red Cross.

"Airplanes bearing doctors, nurses, relief workers, blood plasma, morphine, and even embalming fluid jammed the air over the small airport. From Washington, Chief of Staff Dwight D. Eisenhower offered Governor Beauford Jester any aid with the power of the United States Army. At San Antonio, General Jonathan Wainwright, commanding general of the Fourth Army, sent thirty-one planes and a convoy of forty trucks, ambulances, and mobile mess kitchens. General Wainwright, himself, as did the Governor of Texas, flew immediately to the scene of the disaster. . . . The Eighth Naval District headquarters at New Orleans rushed aid by plane from New Orleans, Orange, Dallas, and Corpus Christi naval installations. . . . The Coast Guard ordered cutters in the vicinity to aid waterfront work. Commercial airlines announced they were sending special planes bearing blood plasma, penicillin, and workers. In Washington, the Federal Communications Commission authorized an emergency amateur network to help handle communications."

For the next several hours, Kathryne will receive friends, neighbors, and relatives who take turns pacing around the house, puttering in the kitchen, making small talk, and distracting the children. Occasionally she flips the radio back on:

"There were five hundred persons watching attempts to put out the blaze aboard the ship when the blast occurred . . . bodies flying through the air in all directions. The blast lifted firemen and their equipment—engines and all—like so many toys. . . .

"It is reported from Chicago today that ammonium nitrate, the inflammable which touched off the Texas City Disaster, is another of man's two-faced friends. In peace, it adds fertility to the soil. In war, it is one of the most lethal of high explosives. When exposed in large quantities to heat, it produces one of the most devastating explosives ever developed by man."

The radio announcer then says he is going to read another updated but still tentative list of the dead.

Kathryne slumps in her chair. She begs for the radio to be shut off.

Henry Baumgartner, the Fire Chief

Twelve-year-old Harold Baumgartner is looking at his dad as the fire chief strides straight into the heart of the smoke. His father is hustling across the deck of the *Grandcamp* and reaching the No. 5 hold.

Twelve of his father's volunteers are just over his shoulders, following their chief, waiting for his orders. His father has beaten back dozens of big fires for years. And now he is staring down at six of his other men—they are below him, on the dock, futilely trying to hose the ship with bursts of water. The entire ship is hot to the touch; the heat is moving in dancing black bands from where the anchor disappears in the boiling water and straight up to the helm.

The instant his men on the dock shoot a column of water, it is like slamming into a wall. Instantly, the water turns to mist.

The French crew has abandoned ship.

The deck is visibly breathing in and out. Harold can still see his grim-faced father and his men in the middle of the whistling, speeding orange smoke—steadying themselves and standing directly on top of fifty thousand bags of flaming ammonium nitrate.

There is a cracking noise rising from somewhere in the lower holds. The ammunition in the No. 5 hold underneath his father is starting to explode . . . and then, from the soul of the ship, there is the apocalyptic, bellowing detonation, the one that people will testify is just like the atomic bombs raging over Japan.

In a fraction of time, it rises through the exact center of the Liberty ship, exactly where Harold Baumgartner last sees his father.

The Priest

As the *Grandcamp* explodes, Roach hears the roar, and then there is the perfect silence. His eardrums are perforated. He is on his back. Everything he is wearing has been blown off his body, except for the blessed medal of the Virgin Mary around his neck. Then the concussion comes quickly, and it is an invisible force, as if gravity has been intensified to the point where it is impossible to raise your chest a fraction of an inch. The concussion flattens hard against his upper body like blunt trauma to his heart. Roach staggers to his feet and is just crossing The Company tracks. As he sets foot on the rails, a boxcar loaded with chemicals heaves into the air and explodes.

Roach continues walking.

Somehow, he finds his way from the waterfront, moving through El Barrio, across Texas Avenue, and then five blocks up Sixth Street. He is going to City Hall. It is where he always goes when he needs answers or demands something be done.

He moves like a zombie through the smoke and the carpet of dead bodies lining the streets leading west from the waterfront. At City Hall, he finally collapses among the fifty-five other people who have already fallen to the ground.

Under the trees, an instant, improvised triage station is developing. Teenagers from Central High are sprinting to Agee's Drug Store, ignoring the expensive bracelets and watches scattered in front of King's Jewelers and picking out bandages from the rubble. The city police cars are slamming to a halt, and bodies are being hauled out of the backseats and lowered onto the grass. There are milk trucks, school buses, and family cars converging on City Hall and the Emken-Linton Funeral Home two blocks south. The drivers have come downtown, looking for news, for relatives, and they've found their vehicles commandeered for rescue trips to the docks and then back to the clinics and hospitals.

Roach is being lifted and moved. He is taken to the city auditorium, and then he is brought outside so that somebody can speed him to the hospital in Galveston.

Curtis Trahan finishes combing Roach's hair. The priest is gently lowered into a seat on the bus that will take him over the causeway and toward the Merchant Marine Hospital in Galveston.

The funeral home operator who helps Roach onto the bus tells people: *"Roach didn't say a word. He was beyond talking."*

AT THE BEELER-MANSKE Clinic, military veterans say it looks like something from the front lines of World War II. There is no water and no electricity, and the windows are gone. Next door, at Agee's Drug Store, parents sift through the rubble, looking for something that they can take back inside to the clinic to stop their children from bleeding. A nurse named Dorothy Wilson is literally walking in sticky blood. A woman is waiting for attention; her back is completely blown off; there is no flesh, and you can see her veins and arteries pulsing. She dies as Wilson and the other clinic staff members look out into the front yard and see people lying on the lawn. Wilson knows they are also dying.

At the Danforth Clinic the floor is coated with glass, insulation, and blood. A nurse listens as a man walks in the door, carrying his wide-eyed and bloodied wife.

"Put your wife on the table, sir."

The man, dazed, replies: *"I can't. I can't let go of her. I'm holding her insides in."*

The woman is finally lowered to a table, and she has a brief moment of coherence when she recognizes a local minister, Rev. Roland Hood.

"It's too much. I just can't take it, Brother Hood."

The clinic staff watches as she passes away.

Texas City cannot handle, absorb, the thousands of wounded people. In Houston and Galveston—at St. Mary's Hospital, Fort Crockett Hospital, Merchant Marine Hospital, John Sealy Hospital, and St. Joseph's Hospital—emergency personnel are racing to the front doors, waiting to receive the first victims.

On Sixth Street, the city bus pulls away from the Danforth Clinic with two dozen wounded passengers onboard and the bus driver aiming for the Galveston hospitals. The driver heads south, circling through the burning refineries. Seven-year-old Foy Smith is onboard and staring at the nightmare, at the way the storage tanks now seem to be melting and spilling out onto the road in front of the bus. His mother is next to him. She is bleeding from cuts on her face, losing blood at a rapid rate, and her nose seems almost severed. In front of the bus, the road is blocked by abandoned vehicles and burning piles of debris. The driver jams on the brakes and yells over his shoulder to his wounded passengers: *"I'm putting this to a vote. I'm not sure if I can get down the road to Galveston. I can turn this thing around and take us up to Houston instead—or I can try to barrel through."*

Galveston is closer, only ten to twelve minutes away. Smith prays that the people will decide to go to Galveston. If they go to Houston, his mother will bleed to death. In the middle of the bombed-out road, a vote is taken and the driver is told to brave it through the heart of the destruction. . . . *Take us to Galveston.*

Longshoreman Jim Trotter can see nurses and technicians waiting outside John Sealy Hospital with gurneys. Trotter is in a blanket, slipping into shock. He is trying to piece together his memories: *He had been standing next to Sandberg on the docks. . . . He was flying end over end. . . . He saw himself suspended somewhere that was soft white. . . . Then he was underwater, his left leg barely attached.* As Trotter fights to remember what happened, nurses at John Sealy are helping him onto a gurney. They make way for a priest who looks down at the longshoreman. Trotter hears the priest begin to read the last rites, and Trotter gamely raises his head: *"Wait a minute . . . I'm not Catholic, and I'm not near about dead."*

At the same time, Joseph Vasquez, a longshoreman friend of Julio Luna's, is finally wheeled into St. Joseph's Hospital. He is lying on a cot in a hallway and the doctor is staring at him.

"*Is he a Dago? Take him to 223, the morgue; he's gonna die soon.*"

The nurse wheels Vasquez into the morgue and covers him with a sheet. A cleaning woman comes into the room filled with dead bodies. Under the sheet, Vasquez asks for someone to please give him a glass of water. She is scared; she tells the doctor there is someone alive in there. He says: "*Go ahead, give him all the water he wants. He's gonna die.*"

For the next seven hours, Vasquez lies in the morgue until doctors finally see that he is simply not going to die. He is moved out.

One thousand seven hundred and eighty-four people from Texas City are going to be admitted to twenty-one area hospitals. Calls are being placed for emergency blood shipments. And calls are being made to military officials for massive amounts of antitoxins that might help the hundreds of people who were immersed in the boiling chemical stew inside the north slip or who had lingered for hours alongside spewing petrochemical tanks.

Beginning around 10:00 A.M. in the hospital hallways, there is something else. A pinging noise is coming from each operating room. Doctors are extracting thousands and thousands of glass beads, shards, and chunks from patients and dropping them into metal buckets. As doctors prepare their patient charts, they notice a trend.

There are several victims who bear no obvious physical injuries. At Sealy Hospital, physicians are noting in their records that some patients "*. . . simply died from irreversible shock.*"

ADMITTING NURSES ARE double-checking.

There seem to be an inordinate number of black patients—the victims who need to be wheeled over to the segregated "colored" wings of the clinics and hospitals.

El Barrio and The Bottom have been hammered more than any other neighborhood in Texas City. Roach had always felt that those neighborhoods were like separate worlds in orbit around Texas City—and now those neighborhoods are enduring the disaster in outsized ways. Roach had talked to Curtis and Ceary Johnson about a color-blind society in Houston. He had argued, in big type, on the pages of the *Texas City Sun*, for the blacks and Mexican Americans to be treated just like the whites. And now, everywhere, people are trading skin colors.

Curtis has seen the irony as he looks out the shattered windows of City Hall. Even Roach looked black. Fleeing north of Texas Avenue are dozens of white people coated black with the oil and molasses. As they run along Texas Avenue, there are some black residents from The Bottom fighting their way out of collapsed storefronts. They are coated white from the cascading asbestos.

Outside Dave's Army-Navy Store, seven-year-old Freddy Vasquez sees women and men who he thinks are from El Barrio—but they have all been turned into black people. Ernestine Garza is screaming for her mother, who is running right in front of her. Her mother turns but doesn't slow down. Ernestine realizes that everyone running is now black—everyone looks the same. Calvin Martin, who had watched his friend at Monsanto go sailing from the smoke shack and thirty feet up in the air, is trying to focus on someone he thinks he knows. For some reason he looks black. "He looked like a tar baby." A railroad car loader for The Company, Gordon Holt, knows what he looks like and he can't wait until "I [am] at last clean and look . . . to be among the white race again."

Monsanto truck driver Julia Turner, the woman who had all her clothes blown off and then ran through town praising Jesus, is picked up, wrapped in a blanket, and taken to Fort Crockett Hospital. She is immediately put in the Negro ward. She will stay there for several days until a black friend recognizes her and has doctors relocate her to the white section. Emory Fletcher, whose son found him hanging on to a smoking boxcar and was able to identify him only by his gold teeth, will be in the colored ward in Houston for three days until his family finds him.

H. O. Wray, who works in The Company offices alongside Mikeska and Sandberg, says: *"I thought it was the Resurrection Day. . . . I wondered why these and other corpses didn't get up. . . . It struck me as rather peculiar, knowing that I was a white man, that I would be in among so many colored people. I did not realize until several hours afterward that I was the same color myself."*

———

CEARY JOHNSON HAS come to the same conclusion as H. O. Wray.

It is Judgment Day.

As the tidal wave thunders down onto Frank's Café, Johnson can see it coming. He is staring right into it. It knocks straight into his chest, lifts him off the ground, and carries him two hundred feet farther inland. In front of him, he sees the eighteen people from Frank's Café being pulled by the retreating tidal wave into the chemical cauldron in the harbor. He watches the wave take Curley Hanson and all the other men who are his neighbors and friends. He can see their eyes as their hands flail in the brown water. It is happening in slow motion, like a stack of those cards that you flip, one after the other, to simulate movement. Each card flips incredibly slowly, and each one moves his friends closer to the bay where the *Grandcamp* had been. The wave is dotted with faces frozen in fear. Ceary knows his friends are already dying. He tries to stand, and there is fire everywhere. The first words in his mind are the same ones that had come out of Bill Roach's mouth this morning when he threw his cigarette down and decided to come to the waterfront:

"This is it."

Johnson begins to lift his feet through the muck.

"The world is on fire. It's Judgment Day. This is the end of the world. Here it is."

He begins to run, not sure where he is headed. Suddenly he is alongside Rankin Dewalt, the gun-toting security guard for The Company whose brother, Commissioner L. C. Dewalt—the former chief of police—went into The Bottom and fired five bullets into a black man.

Johnson and Dewalt silently follow each other, looking for familiar reference points. At 4th Avenue, Dewalt veers north to the white neighborhood and Johnson heads west to the Mexican and black neighborhoods.

As Johnson begins to fix his location, he is staggered.

Every shotgun shack and wood bungalow in El Barrio is either knocked down, blown out, on fire, or cracked. Most of the already flimsy homes in The Bottom are as bad. Our Lady of the Snows

Church, the rallying point for Texas City's Mexican American community, is in danger of collapsing. Out front, the only thing unharmed is the statue of Our Lady; an elderly woman, maybe in her eighties, is steadfastly bowed down in front of it.

At Norris's Café on Sixth Street South, people are lying down and dying at the front door. More people head to the café. There is no place else to go in The Bottom. Johnson crosses Sixth, and ahead of him he can see that the Booker T. Washington School is sagging.

Inside, first-grade teacher Evelyn McClure is struggling up, trying to remain calm and staring at James Lee, who is passing out twenty-five cartons of milk and whose forehead has now been pierced by a piece of steel. Just a few minutes earlier, the children had finished their usual morning routine: raising the flag, singing the "Star Spangled Banner," reciting the Pledge of Allegiance, saying the Lord's Prayer, and finishing with the Negro National Anthem.

Now, McClure is lifting each of her twenty-four children by the pants and dress sleeves and throwing them out the holes where the windows had been. She is screaming for all of them to run to their parents. Behind the school, on the baseball field where sixth-grader Harold Adams and nine of his friends had been doing their best Jackie Robinson imitations, only Adams is there now. His friends tried to get him to go to the docks, and now he is staring at the smoke from the waterfront; he knows that all nine children have just died.

Down the block from the school, people are running to the house of Rev. F. M. Johnson. He is just about to step outside to drive over to see the fire when the explosion begins to knock down the windows. Reflexively, Johnson is ducking. Maybe someone thinks he's gotten too uppity. Maybe someone is trying to kill him.

Either a bomb has gone off or someone is shooting at me.

Just before the small crowd reaches his house, a piece of steel falls through his back door. His neighbors want him to drive them somewhere, to tell them what to do, where to go.

Johnson still isn't fully dressed. The preacher runs into his yard, opens his car doors, and tells anyone who can fit in to get in. He speeds to Galveston and then speeds back and forth for the rest of the day, the blood soaking deeper into the upholstery in his '41 Chevrolet; some people hang on to his car's fender, and some simply drape themselves across the hood.

Ceary Johnson passes the preacher's house and is running to his house on Seventh Street, looking for his wife. Two hundred fifty miles away in Jeanerette, Louisiana, Ceary's father has felt the blast and is wondering what the hell it's all about. His son is mentally preparing a list of the families that he knows he is going to have to visit. Ceary's a gang boss, someone his men have relied on.

Now he is going to have to figure out a way to tell these families their husbands are dead. He wonders if he should tell them exactly how they looked when he saw them die.

As he runs to his house, he is remembering someone else he had seen at the waterfront.

Father Bill Roach.

IN MERCHANT MARINE Hospital, people are making way for Galveston bishop Christopher Byrne.

He is an august man, someone who has just been in contact with Pope Pius XII about the fact that this part of Texas has one of the oldest and richest Catholic traditions in the United States. His diocese is sometimes called "the mother diocese of Texas." Just two weeks ago, the Vatican granted Byrne a special benediction honoring the hundredth anniversary of the diocese, and it empowered him to offer special blessings to the people in Texas City. Four days from now, he was scheduled to oversee the dramatic opening of the first full-scale hospital in Texas to serve Negroes. It is called St. Elizabeth's, and it is on Lyons Avenue, considered by some to be Texas's version of 125th Street in Harlem.

The hospital has been designed, orchestrated, and willed into existence by John and Bill Roach.

This morning's black newspapers in Texas feature stories and congratulatory messages—saluting the Roach twins for their work getting the revolutionary hospital funded, built, and opened.

By now, within hours of the explosion, Byrne has decided to cancel his special blessing ceremony for Texas City—and also to cancel the dedication ceremonies for the Roaches' hospital for Negroes.

Byrne is pressing down the crowded, noisy, and impossibly frantic Marine Hospital hallway, pausing to deliver the last rites to people reaching up from their bloody cots. Finally, he is ushered into Roach's operating room. The doctors are still reaching the same evaluation that Curtis Trahan and others had come to when they saw Roach in the minutes after the *Grandcamp* explosion.

He is beyond talking.

Byrne sits alongside Roach, clasps his hand, and tries to summon a hint of recognition from the priest. Roach, with the tattoos of oil scrubbed from his skin, looks exactly like the impudent wiseacre who had barnstormed into the Bishop's Palace almost a decade ago. His skin, rubbed with alcohol, is ruddy again. Byrne sees what everyone has seen. There are no wounds; there are no scars. But Roach's eyes are wide open, as if he has encountered something incomprehensible that has frozen him in perpetual shock.

Back when he first met the Roach boys, Byrne didn't know whether he could trust them. He suspected that they were full of it. That they were blustering, wild Irishmen who might be too unfocused to do any good. That they might be more trouble than they were worth. Byrne had laughed at their old Ford, laughed at the way the Roach boys had set him up by sending only one brother in to bargain for admission to the priesthood—and then how that brother suddenly mentioned that there was another brother . . . *just like me* . . . waiting in the car. Byrne had decided to roll the dice, to take a chance, to see if the Roach boys were tapped in to some new way of doing things.

And then he had held his breath for several years as critics and enemies said the Roach boys were dangerously close to no longer being priests—they were becoming socialists, communists even, people who had forgotten that they were supposed to concentrate on spiritual matters instead of delivering medical services, food, water, and homes to poor black people.

Byrne never chastised them. The diocese files on the Roach boys contain no negative reports filed by Byrne. He had let Bill Roach, especially, run free. And the twins seemed to be finally converging on something good. Bill Roach had convinced the mayor of Texas City to tackle the problems of the blacks and Mexican Americans—and in

Galveston a hospital was being built where nothing like it had existed before.

Now Byrne has his hands on Roach, and he begins to pray. He knows that there are clergymen and parishioners all along the Gulf Coast who have come to believe that Bill Roach is a saint in their midst.

Byrne begins to say the last rites.

While Byrne prays, Johnny Roach is doing what the Roach boys do. He is bluffing, blustering, talking his way past the police lines that have sprung up on the outskirts of the Texas City waterfront. He drove there as soon as he heard the news about the explosion. He is hell-bent on finding his brother. Johnny is beyond delirious. He is arguing, cajoling the Texas City police officers; finally they are letting him through. Outside his car window it is like scenes from hell. His brother had had visions.

Bill was right. Blood is running in the streets of Texas City.

As he drives, some wounded people turn toward the face they see in the slow-moving car. They start to the car, calling after it, waving for it to slow down. They are sure they know who is inside. It looks exactly like Father Bill Roach.

For weeks, people will refuse to believe that Bill Roach is dead. There will be reports that he has been seen at City Hall, talking earnestly with Curtis Trahan. There will be reports that he is going door to door in El Barrio and The Bottom.

With Bishop Byrne holding him in his arms, Bill Roach dies in Marine Hospital.

The Second Explosion

CAPTAIN MOSLEY PETERMANN, the *High Flyer* skipper, sees sheets of fire spiraling from the *Grandcamp*. He orders the piercing emergency alarm to sound, scrambles his men to their fire battle stations, and screams for all the hatches to be sealed.

The lower holds of his 438-foot vessel are brimming with 961 tons of ammonium nitrate.

The *High Flyer* and the other Liberty ship in port, the *Wilson B. Keene*, are moored in the main slip. They are only six hundred feet from the *Grandcamp* in the north slip.

When the first explosion rolls out, it instantly rips the four-block-long *High Flyer* clean from its moorings, as if a thread is being yanked from a spool. Dragging its two-ton anchor chain, it speeds sideways across the slip and cracks into the hull of the *Wilson B. Keene*. The hatch covers are immediately blown wide open. The portside superstructure of the oceangoing freighter is crumpled like a ball of paper. Pearl Harbor veterans say it is like that day in 1941 when tethered warships looked as if they were bucking and alive, trying to ride out the fiery chaos.

Petermann's sailors stumble from cabins and belowdecks, blood on their uniforms and coughing inside shrouds of burning sulfur. As the refinery tanks begin to blow up one by one along the waterfront, the water in the bay roars, surges, and the two huge ships in the main slip scrape against each other and rock horribly up and down, as if they are colliding in thirty-foot-high mid-Atlantic storm seas.

Captain Petermann orders the anchor raised, but his engineers yell back that the ship's windlass is destroyed.

At 10:00 A.M., Petermann gives the order to abandon ship.

The sailors jump onto the deck of the *Wilson B. Keene*, scramble portside, grab rope ladders, and slide down them somewhere into the fire and yellow chemical clouds.

CURTIS IS RECEIVING reports that the twenty-seven members of Henry Baumgartner's volunteer fire department most likely were killed instantly.

They have vanished at ground zero of the explosion.

It is the worst fire department tragedy in the twentieth century. More firefighters have died in one single instant than have ever died in any other previous fire in the nation.

Since he first climbed over and through the debris at City Hall, Curtis has been flinging out large-scale duties. He has already told twenty-eight-year-old John Hill—a neighbor for God's sake . . . somebody he plays bridge with—that he wants him to handle the army, when one of the immense heroes of World War II, Fourth Army general Jonathan Wainwright, arrives.

When Bataan fell unconditionally in one of this country's most devastating military campaigns, Wainwright and seventy-six thousand Allied prisoners of war were forced to walk sixty miles to a Japanese camp. More than five thousand American men died on the march. After the war, Wainwright was awarded the Medal of Honor.

For the next three hours, waiting for Wainwright, Curtis is racing to delegate mammoth priorities to anybody he thinks is a remotely responsible human being.

Local physician Fred Quinn is told to call hospitals, ambulances, and anyone else he can come up with. Fred Dowdy, a fireman who happened to be out of town when the explosion occurred, is back, and he has been told to coordinate the dozen fire departments converging on Texas City. Police Chief Willie Ladish, who has only seventeen people on his staff, is already in conference with the Texas Rangers, the Houston Police Department, and, especially, the FBI.

That morning, before the blast, Ladish's biggest worry had been dispatching cars to unravel traffic jams and logging his men off for an hour or two as they stopped inside Clark's Department Store.

By 2:00 P.M., teams of Houston-based FBI agents are asking him about interviewing eyewitnesses. Orders have come directly from the office of J. Edgar Hoover to investigate the possibility that a terrorist's bomb had been dropped or planted on the waterfront.

America is locked in a Cold War, as President Truman's adviser Bernard Baruch had just announced to the world last night. Hoover wants Texas City canvassed immediately. It could be the beginning of a massive terrorism plan to target vulnerable, strategic American cities. It could be the beginning of a widespread, carefully coordinated plan to spread destruction and panic on American soil.

The agents especially want to talk to the French crew—members of the crew reportedly have Communist Party cards in their immigration papers. The National Maritime Union, the International Longshoreman's Union, the Oil and Chemical Workers Union—they are all busy, active, feisty on the Gulf Coast. The FBI agents know, already, that there are some longshoremen who had also belonged to the Communist Party, the Socialist Party, the Industrial Workers of the World.

If it was a terrorist plot, there could be more saboteurs roaming around the Gulf Coast. Amid the chaos and rubble in Texas City, it is easy to spot clean-suited FBI agents moving in teams of two, passing through police barricades and buttonholing refugees running away from the dock area.

Curtis can see the FBI agents outside and inside City Hall, moving on the periphery of everything, and always, always, taking notes.

Curtis knows that Hill is right.

I've got to get this damn thing more organized.

A reporter is at City Hall, running alongside him and shouting questions.

"You were in the Battle of the Bulge; you got a Purple Heart when a buzz bomb almost blew your leg off. How does this compare?"

Curtis stops.

"No buzz bomb could ever compare with what happened here."

CURTIS HAS ASKED for a cot to be delivered to his barely recognizable office. He'll be sleeping here. He sits at the cot's edge and keeps coming back to the one elemental, cruel task: In the rush to evacuate victims, no one has been exactly certain about where to bring the hundreds and hundreds of bodies.

Two blocks to the south, the Emken-Linton Funeral Home is one place where people are unloading corpses. At McGar's Sinclair station, Bob McGar has cleared out his garage and told people driving by that they can bring bodies there. Some people are bringing bodies to the high school gymnasium. Bodies are being left outside Norris's Café in The Bottom. Funeral homes have dispatched cars from Galveston, Houston, and every small town along the coast and the bay—and they are picking up bodies almost everywhere in El Barrio, The Bottom, and near the ruined, still-blazing Monsanto plant.

Curtis is beginning to absorb the extent of the cruel irony.

He's been mayor only a year.

The communities he and Bill Roach had just won unprecedented victories for are devastated.

Only six days ago he and Roach had forced the big companies to join Texas City, to pay taxes, to fund schools and streets and sidewalks and parks. Just six days ago he and Roach had also ramrodded the plan to render instant aid to the poorest part of Texas City—to Ceary Johnson, Florencio Jasso, Rev. F. M. Johnson, and everyone else who lived there.

Now those parts of town that he and Roach were hell-bent on rescuing are virtually erased.

El Barrio is in tatters. The community's church, Our Lady of the Snows, is an aching metaphor. It is caved in, useless, beyond repair. Row after row of houses in The Bottom, including the one that belonged to the woman who baby-sat his children, are mangled.

It is, in a very real way, as if the communities have themselves become corpses.

Curtis and Edna had always told each other that this was it—that they were never going to leave Texas City. He had never been more convinced of it than a few days ago when he had come home that night, flushed with victory and righteousness over winning the annexation battle. He had beaten whatever the hell the old racist vanguard of Texas City was desperately trying to hold on to.

Now Texas City, the Texas City he had envisioned, is virtually dead.

Curtis begins to draft the first Emergency Order of his administration. It is as if he is trying to hold together the core of the city, not to let anyone intrude on it, not to let anyone demean it, literally to keep the people of Texas City whole:

> *"By virtue of the authority vested in me by the Charter of the City of Texas City, I, Mayor J. C. Trahan, hereby establish a commission to be known as the DEAD BODY COMMISSION. The purpose of this body will be to regulate the moving, removing, embalming, identifying, and releasing of bodies of persons recovered within the limits of the City of Texas City and to prevent a recurrence of the unfortunate instances when bodies have been lost prior to identification and without permission of relatives. . . .*
>
> *"The bodies of all persons found or to be found within City Limits shall be removed to the building owned by McGar Motor Service at 618 Fifth Avenue North in Texas City.*
>
> *"No dead body shall be removed from the place where it is found except through and by vehicles operated by authorized personnel of the United States Army. . . .*
>
> *"No dead body shall be removed from the corporate limits of the City of Texas City without written permission of the Dead Body Commission. . . .*
>
> *"Any person violating any order of the Dead Body Commission issued pursuant to these rules shall be arrested immediately. . . . ."*

Now THE BATAAN DEATH MARCH general is at the head of his convoy, moving into Texas City and approaching City Hall.

It is 5:30 P.M.

John Hill is out front, waiting for them and watching the Red Cross volunteers and crews man their tents, tables, and cots.

Within thirty minutes of the explosion, Red Cross volunteers from

Galveston had begun arriving in Texas City. By chance, that morning, the Galveston chapter had taken possession of 210,000 surgical dressings from government war surplus supplies. The first doctors, nurses, and medical students to arrive are there within the hour, dispatched from the three hospitals in Galveston. As the first wave of fifty-five medical personnel is cleared through the police barricades, they are dropped off at the three clinics, at City Hall, and at the Red Cross tent that has just gone up on Sixth Street. As the first radio reports filter across Texas and Louisiana, emergency personnel—fire crews, rescue workers, physicians, security details—are ordered to Texas City from Dallas, Fort Worth, San Antonio, Port Arthur, Orange, Corpus Christi, Austin, Baton Rouge, Shreveport, and New Orleans.

Hill stands in front of City Hall and looks toward the confusion and the chaos that trails up from the Mainland Drug Store, Agee's Drug Store, and the tiny Texas City National Bank. There is a late-model Buick, painted army olive drab, moving along Sixth Street. A general's flag flutters from the front fender, and it is leading miles of military trucks, ambulances, and Jeeps.

The Buick comes to a halt, the door opens, and a figure in full uniform steps out and stands in front of Hill.

"I'm General Jonathan Wainwright. Mr. Hill, the Fourth Army is here, and we are ready to take over."

Hill replies: "General, so far as I know, Governor Jester has not declared martial law, and, until he does, I assume I am going to be in charge."

Wainwright, who had lived through the Bataan Death March and spent thirty straight days on the rocky jungle island of Corregidor with ten thousand soldiers locked in a losing and hopeless battle against a Japanese onslaught, measures the twenty-eight-year-old for a second:

"Mr. Hill, you are absolutely right. What do you want the Fourth Army to do?"

Hill turns to walk inside and toward a city map hastily pinned to one of the still-sturdy walls in City Hall: "General, come with me."

The ex–baseball hotshot leads the full general, the Medal of Honor leader of the Fourth Army, to the map and begins pointing to places where some battle-hardened troops need to be stationed.

There are reports that some have already crept into Texas City. Mysterious skiffs have been spotted just offshore, some of them lingering over the bodies floating in the bay. Rowboats have been tucking onto the sandbars; men have been jumping off and scavenging through the charred remains on the waterfront.

FIFTEEN MINUTES AFTER Fourth Army general Wainwright arrives, a radio reporter for KPRC of Houston is going on the air with a live national report: *"In the temporary morgue one hundred yards from our broadcast point, there are an estimated two hundred bodies being identified and embalmed. Once the bodies are identified, they are moved several blocks away to the floor of the high school gymnasium. . . . At this point of our broadcast, hearses are lined up as far as two blocks away, waiting to transport the dead."*

At 6:20 P.M., soldiers from the Fourth Army units are strapping on chemical warfare gas masks and driving transport trucks loaded with stretchers to the waterfront. The children still roaming along lower Sixth Street and across Texas Avenue are staring at the camouflaged Jeeps filled with men in heavy boots, work gloves, long-sleeved shirts, and the thick-lensed gas masks.

The Fourth Army Jeeps backtrack through the endless chemical ponds, some of the small bodies of water afire, and they marshal out to where the Monsanto plant, Frank's Café, and El Barrio once stood. Behind them there are rows of bulldozers rattling in the same direction with more soldiers onboard, to begin digging through the collapsed buildings, warehouses, and homes.

At 6:25 P.M. Curtis hastily calls his police chief.

There are too many little children lost and wandering everywhere. I want any child who's found alive to be taken off the streets right now.

A BLOCK AWAY from where Curtis Trahan is overseeing operations, Henry Baumgartner's wife, Christine, is on the phone at her house on Seventh Street.

She is talking with her friend Mattie Westmoreland. Mattie is asking if Christine knows anything, if she has any word at all. Mattie's husband, Jack, is a volunteer fireman. He had raced to the docks at 8:30 A.M. when the general alarm sounded. Christine simply says: *The entire fire department is gone.*

Then Christine says something strange to her friend: *I've been clean-ing all the glass out of my house. I want it clean when Henry comes home. You should do the same thing, Mattie.*

At Mikeska's home on Ninth Avenue, his daughter is going back and forth between the phone and the front door. Dozens of employees from The Company are stopping by. There are rumors: Her father is in a hospital bed in Galveston. Someone has seen him walking by City Hall. He was with Sandberg. He was helping people escape what was left of The Company's dockside offices.

People put on stretchers by soldiers from the Fourth Army and then raced through the fires alongside Monsanto and Pier O are telling their rescuers they have spotted someone else.

They've seen a Catholic priest. He has been delivering the last rites, going from one body to the next.

They are insisting that it looks exactly like Father Bill Roach.

At 7:30 P.M. Trahan, Hill, and Sandberg are at City Hall.

A weary rescue crew is just back from the docks. They've pulled out. There's too much smoke, still too many unpredictable fires. The entire bay looks as if it is under one perfect black cloud. One of the rescue workers has an urgent message for Trahan and Sandberg: *"There's another ship down there with that dynamite on it that's on fire."*

Sandberg has no idea what the hell he is talking about.

"What do you mean, dynamite?"

The rescue worker, almost defensive, replies: *"I know there is, because here's a fellow that helped put ammonium nitrate on her."*

That morning, before the *Grandcamp* blew up, Sandberg had told the owners of the *High Flyer* that she was safe.

Now more rescue workers are still dumping bodies on the front lawn of City Hall and yelling through the broken windows that the *High Flyer* sailors say their ship is next.

Trahan tells Sandberg that he's giving him the *High Flyer* situation.

Mikeska and Sandberg run the damned waterfront. The Company owns and operates the damned docks. They sure as hell should know what each goddamn ship has in each and every hold. Even police chief Willie Ladish has been going over the inventory of the *High Flyer* with

Sandberg. He is with The Company. He has to know, should know, that there is ammonium nitrate on the *High Flyer*.

Trahan wants Sandberg to deal with the *High Flyer*.

It's yours.

Sandberg calls the *High Flyer*'s owners back. They want their ship moved—just the way that Father Bill Roach had wanted the *Grandcamp* moved.

Sandberg doesn't know what to say. He talks to John Hill—he's a chemical engineer; he works at Union Carbide—maybe he'll know what the hell is going to happen to the *High Flyer*. Hill says he'll check with the Union Carbide chemists.

For decades, Hill and Sandberg will differ on exactly what is said between the two men. Sandberg remembers it this way:

"They did say this, that if the fertilizer or ammonium nitrate caught on fire and burned a sufficient length of time that it reached a point of explosion, that it would just have the effect of a brief explosion, and that as long as men were not directly in the ship they saw no danger."

After Sandberg relays this to the owners of the *High Flyer*, he hangs up.

There is a persistent reporter milling around City Hall. He's been hearing the same scary things as everyone else. You can't help but hear them. The reporter corners Sandberg: *Is that other ship going to blow? Is there going to be another explosion like the one this morning?*

Sandberg says: *"No, there is no danger of any other ships blowing up."*

By 8:00 P.M., anyway, several radio stations are beginning to report that there is going to be another explosion and that it is going to be as devastating as the first one that occurred that morning. A truck with no doors, two flat tires, and a megaphone hooked to the roof is racing along Sixth Street, along Texas Avenue, north on Bay Street; it is blaring a message:

"You must evacuate Texas City immediately. Repeat. You must leave Texas City immediately. Leave your homes now. Do not stay inside your homes. You must evacuate Texas City immediately."

Just after 8:00 P.M., the telephone operators in Texas City are being handed their own Fourth Army gas masks.

All of the women employees are sent home.

General Wainwright has been given a top-priority escort past the barricades in front of the demolished docks. From his Jeep, he can hear screams coming from inside the still-burning sections of the Monsanto plant. The juxtapositions are not unlike the ones he saw overseas during the war—there is, in the middle of the bombed-out buildings and homes, a small group of cows placidly chewing on a square of unburned grass. Wainwright has been witness to and a participant in the most devastating moments from World War II. In Texas City, the normally loquacious commanding general of the U.S. Fourth Army has run out of words. He turns to a reporter who has been trailing him: *"I have never seen a greater tragedy in all my experiences."*

There is little argument from the veteran war correspondents who have been sent by the major news outlets to cover the story in Texas City.

Hal Boyle, a front-line World War II correspondent for the Associated Press, is filing his firsthand account and comparing it to atomic warfare: *"In four years of war coverage, I have seen no concentrated devastation so utter, except Nagasaki, Japan, victim of the second atom bomb, as presented today by flaming Texas City.*

"The damage along the waterfront exceeds in intensity that inflicted on Bari, Italy, in the fall of 1943 when German bombers hit that port and seventeen vessels went up in flames, including three ammunition ships, the titanic explosions of which killed hundreds of American troops and Italian civilians."

By 8:00 P.M., police and soldiers are in the streets, moving in teams and edging people out of buildings. A longshoreman, pounding plywood over what's left of his window and door frames, peeks over his shoulder to see one of the Texas City policemen glaring at him, waiting for him to finish and get the hell on the highway to Houston.

There are more sound trucks rattling down the streets:

"ANOTHER EXPLOSION IS IMMINENT! GET OUT OF TEXAS CITY NOW!"

ON THE DOCKS of the Bay Towing Company in Galveston, sailors are being given a briefing: *There is ammonium nitrate onboard the* High Flyer. *We think it's on fire. It's got at least eight hundred tons of the same explosive that blew up the* Grandcamp.

Now the sailors are being asked for a show of hands.

Who volunteers to sail to Texas City and move the High Flyer *out to deep water?*

One by one, the hands go up until there are four crews ready.

At 8:30 P.M., the tugboats *Guyton, Albatross, Clark*, and *Miraflores* begin to sweep through the night, past Bolivar Peninsula and toward Texas City. The northwest winds are blowing in the sailors' faces at fifteen mph, and each gust pushes the black clouds of smoke over the tugs. The sky is still painted black, and there is no glimmer of the usual climbing moonlight. The tugs slap against the shallow, rippled water and begin to line up to make a pass into the harbor area.

As the tugs negotiate the debris and the bodies in the turning basin, it is close to 10:00 P.M.

Ahead of the *Albatross,* the *High Flyer* looks like a living, angry beast pressed against the *Wilson B. Keene*. Angelo Amato, at the helm of the *Albatross,* thinks the damned thing is a "volcano." Thousands of neon red sparks are firing in columns out of the ship's holds. The dock is impossible to see. Timbers, some on fire, are slapping against the tugs. Massive pieces of jagged, glowing metal are sticking straight out of the water.

There is something that sounds like a massive breathing noise, and it is growing louder as the tugs come into the turning basin. The glow of the shower of sparks is set against the urine-colored sulfur smoke from the lower holds of the freighter. The four tugs are bobbing in the bay, and the crews are leaning around their two-way radios and ship-to-shore microphones. Amato is plotting strategy with the skipper on the *Guyton*.

At 10:34 P.M., national radio audiences are hearing live reports on NBC Radio about what is happening in Texas City: *"It is safe to say every house in Texas City, a town of fifteen thousand residents, suffered at least some damage. The business district, about a half mile from the scene of the blasts, is a shambles. A residential area between the downtown sector and the plants that suffered the greatest damage on the waterfront also is all but destroyed. Huge hunks of steel lie in the streets, blown many hundred yards from the ship whose initial explosion caused the conflagration."*

There is no mention of another ship, also loaded with ammonium nitrate.

The *Albatross* and the *Guyton* are going to make the first approach.

As the *Guyton* pulls alongside the *High Flyer* bow, sailors try to run a hook through the big ship's anchor chain. The work is slow because of the fumes. The sailors are wearing gas masks, trying to see through the yellow billows. After a long while, the hook line is attached. It seems secure, but, without warning, it snaps and ricochets off.

The *Albatross* and the *Guyton* reconnoiter.

The skippers decide to put together a squad of men with cutting torches. They are going to board the ship and reach through the dense, sickening sulfur fumes to cut the anchor chain that the *High Flyer* has dragged across the slip after it was ripped from its moorings.

The tugs jockey for position and slide alongside the freighter. At midnight, sailors from the *Albatross* throw themselves onboard, carrying the acetylene torches. Crawling over the sweltering deck, the sparks falling on their backs, they begin to rappel down the side of the bellowing freighter, their work boots tapping just outside the holds where 4 million pounds of sulfur, large tanks of Bunker C fuel oil, and 19,220 bags of smoldering ammonium nitrate are stored next to one another.

Masks on, the sulfur choking their lungs, they lean into it and finally slice the anchor chain just above the waterline.

Five other sailors from the tugs are sprinting to the port side of the *High Flyer*. They are using their blowtorches to shear the tangled masses of steel where the bow of the *High Flyer* has jammed into the stern of the *Wilson B. Keene*.

Finally, the freighter looks clear.

At 12:50 A.M., lines are secured to the tugs. The *Albatross* and the *Guyton* tie together, and their skippers are gunning their engines. The water is frothing, churning as the tugs strain backward. The lines secured to the *High Flyer* are jolting hard and taut.

There is an awful, echoing screech as the tugs pull and the *High Flyer* scrapes away from the *Wilson B. Keene*. The ship moves twenty-five feet into the slip and then another twenty-five feet. And then the massive thing shudders in the water and is locked in place as if something has grabbed it from below and fastened it down.

There is no give, not an inch.

The helpless sailors stare as the *Albatross* and the *Guyton* are coming back to the *High Flyer*, almost as if the giant smoking ship is reeling the two small tugs in.

At 12:55 A.M., heavy black smoke is huffing up from the holds of the *High Flyer*. From the smoke come giant fireballs, hurling and curving out over the sailors on the tugs. There are screams for someone to cut the damned tow lines.

There are still some men onboard the *High Flyer*.

The *Guyton*'s captain elbows his tug straight into the *High Flyer*'s hull, the smoke coming harder and the fireballs popping like car-sized Roman candles. He's looking through the night, trying to see who the hell has been able to slide down the lines and get back on the tug's deck.

With all hands accounted for, the two tugs are escaping the slip, just getting free and into the turning basin, just trying to gain some distance. At six hundred feet out, the inferno heat melts the bulkheads of the *High Flyer*. Its contents—the exact contents for a bomb capable of leveling a city—are merging at the perfect point of maximum explosion.

At 11:30 P.M., while the tug crews are still locked in a battle with the *High Flyer*, Curtis finds John Hill. There seems to be the possibility of the thinnest sliver of order, at least, as night comes on. Curtis hasn't sat down since 10:00 A.M. He has met General Wainwright. The FBI agents want to ask more questions. At McGar's garage, they want to know what else to do—there are too many bodies splayed across the service station already. Curtis sometimes walks the grounds of the City Hall complex—to the city auditorium, to the Red Cross shelters, to the Salvation Army booths, to the Fourth Army field tents. He visits the phone operators. He gives pep talks. He offers updates . . . *no, he didn't know if there was going to be another explosion . . . no, he hadn't heard about gas lines under the schools being in danger of exploding . . . no, he had no final news about Father Bill Roach.*

Now Curtis is ready to plot. At 11:30 P.M., he tells Hill: *"I'm ready to sit down and get things organized."*

The two neighbors walk to the second floor of City Hall and sit in the room normally used for City Court. They have writing tablets and pencils. They stare down at the clean, blank pages, and then they stare at each other.

"Okay, what the hell do we do?"

Gradually, they decide to form committees and appoint their friends, the people they can trust to run them, the people who have already seemed to take charge even if they didn't know they were taking charge. They decide that they want Texas City to oversee as much of its own salvation as possible. Even if it means antagonizing the Fourth Army, the FBI, and anyone else being dispatched by Washington.

Trahan is jotting down names, bouncing names off of Hill. They argue someone's merits. They draw up more lists. Hill says that Trahan shouldn't be spending all his time answering questions from reporters. Texas City needs a spokesperson, somebody who can kill rumors as they slither onto the airwaves. Trahan tells Hill: *"Why don't you do it?"*

Hill thinks for a second and agrees. He has already gone face-to-face with the hero of the Bataan Death March.

As they talk, they are interrupted by an urgent message from Sandberg: The *High Flyer* is blazing.

Hill tells Sandberg he has to order everyone away from the main slip, to spread the message, any possible way, that everyone needs to flee from the area around the *High Flyer*.

Curtis and Hill go back to their notepads, wondering if any committee of local residents can even remotely corral what seems to be the utter destruction of an entire city.

They are alone in the courtroom, and it almost quiet. Across the city, some people are finally getting to sleep inside the shells of their homes. In the coastal town of Kemah, Forrest Walker and his mother have finally closed their eyes in a stranger's house.

It is 1:10 A.M.

ELIZABETH DALEHITE IS at her home on Offat's Bayou, across Pelican Island and the west side of the bay.

The old house, set on the water, is like something airlifted from an old New England fishing village and transplanted into the squalid

Texas Coast. There is a small pier. An oyster reef a few feet from the front door. The huge living room, the first room in the house, is warmed with the burnished items that Henry Dalehite had brought back from his ocean trips, or had been given to him by the endless itinerant sailors who came for drinks, dinner, and stories. It is a pleasantly weather-beaten place. When people come to the bayou and see their home, it is easy to imagine that the residents have a connection to life at sea.

By now, Elizabeth has had time to weigh the fact that she had not joined Henry when he walked away from the car. He had leaned over, looked in the car window, and gently tried to cajole her into coming. She had demurred, feeling a little sleepy, feeling as if she really should adhere to that morning ritual of saying her prayers. Elizabeth watched her husband disappear into the fire. She held the statue of the Blessed Mary as it was decapitated.

There is no news about Henry.

Upstairs, her son, her husband's namesake, is sitting at the edge of his bed. His birthday is coming up; he's about to turn nineteen. She had reached him at school and he had gotten a ride with friends from Austin. It seemed to take forever. Now he is alone in his room, staring out the window at the fires burning in Texas City.

His father had a high school education and had even taken a stab at college; his mother had never had time to graduate. His father loved to bring him adventure books. *The Jungle Book. The Explorers. Huckleberry Finn. Tarzan.* And Henry Jr. had loved working for his father. He loved being out on the low-slung launches that his father called "doodle bugs." When he was fourteen, he had stood alongside his father, leaning into the wind during the hurricane of '43, the one that nobody knew about in advance because the government had squelched the weather forecasts so the German U-boats wouldn't get a leg up. One of those damned launches had snapped free, and he had been on it, hardly able to stand up, and trying to save his father's boat.

There were other moments with his father, incredible moments, like the time they were headed through Louisiana on the Intracoastal Canal and one of his father's boats caught on fire and went down, and his old man had turned to him and asked him to dive under and tie a line onto the thing because they were going to haul it right back out. And he smiled, sometimes, when he thought about his father's pleasure boat—*The Galvez*—the big white party boat that all the rich swells like Frank Sinatra, Phil Harris, and the other Hollywood types liked to sail on for their private parties. There were illegal slot machines onboard *The Galvez*—his old man really didn't have much of a choice because the local branch of the Mafia had insisted—but his father had rolled with it, never stopped laughing and never seemed to cease his moments on the water. His old man also didn't have much choice during the war, when the military simply took the boat over so it could be used to ferry ironworkers to a wartime foundry on Pelican Island.

Henry Sr. was a true sailor, an old salt, and he would have loved it if his only son had decided to take up the life, too—but he seemed just as content, maybe more so, knowing that Henry Jr. was going to be a big lawyer someday.

Now the old man was missing.

The family has a burial plot in Galveston. It is where his mother and father had buried their little baby, Jacqueline, the one who died after only a few weeks.

Henry Jr. hoped to God that they wouldn't have to bury his father, too.

As Henry Jr. sits on the edge of his bed, on the second floor of the sea captain's old house, he is looking out the window and directly at the sky over the Texas City waterfront.

He sees it.

At 1:10 A.M., early on Thursday morning of April 17, a three-thousand-foot tower of fire from the decomposing *High Flyer* soars straight into the night sky, lighting up the Gulf of Mexico and what's left of Texas City just like high noon.

The explosion is greater than the one from the *Grandcamp*.

The President

APRIL 17, 1947

Western Union Telegram

April 17, 1947

To: Mayor J. C. Trahan, Texas City, Texas
From: The President of the United States

I have asked every Government agency to cooperate in relief activity. My heart and the heart of the nation go out in deepest sympathy to the suffering people of Texas City. May God lighten the burden of sorrow which has fallen on the community with such tragic force.

Sincerely,

President Harry S. Truman

The City

❧

As THE sun rises, there are army guards wearing gas masks and brandishing fixed bayonets outside McGar's service station.

There are rumors of deadly chlorine gas in the air, and petrochemical clouds are still blowing out toward the bay, still moving with the winds from the northwest.

A bread truck speeds up—commissioned to carry the dead from the waterfront—and the Fourth Army soldiers flank the vehicle and stare down anyone who tries to get close.

Nurses are moving through the crowd with little vials of ammonia spirits in order to rouse anyone who faints.

A woman is standing on the other side of those pointed bayonets and trying to yell through the soldiers and into the gas station that's now the morgue: *"Do you have any little boys in there?"*

The soldiers finally take pity: *"No. No little boys."*

The mother angles to the side of the building and tries to run up to a window so she can look inside. In the garage, there are hundreds of bodies and limbs lying on black tar paper. There are several bodies with tissue gas, and they are swollen several times the normal size of a human being. Grimly, the embalmers are puncturing the bodies with hypodermics to reduce the swelling.

Before the young mother can look, a soldier runs in front of her and gently pushes her away.

Her face the color of alabaster, she simply retreats into the crowd.

Ceary Johnson watches the crowds. He has walked uptown, up to where blacks are normally not welcome, and he is thinking that he lives in a city on fire.

He is thinking of the Bible verse, trying to summon the words to that passage about water and then the fire next time... *what wasn't destroyed by water ... will be destroyed by the fire next time.*

WITH THE SUNRISE, it's clear that the *High Flyer* has claimed whatever hadn't been taken by the *Grandcamp*.

Houses in El Barrio and The Bottom that somehow refused to succumb entirely the first time around are now just splinters and boards. U.S. Coast Guard cutters, attempting to enter the turning basin to rescue any of the living or deceased flung into the water, are radioing in to mainland base stations to say that the fires are more intense than they were yesterday.

The tail shaft of the *High Flyer* has landed like a javelin in the mud of Rattlesnake Island. Four of the gigantic eighty-thousand-barrel Humble Oil tanks are ignited. A steel Union Carbide tank dissolves into a spreading metallic lake. More than a mile away, the No. 60 tank at Republic Oil is blowing up. A one-ton turbine also attacks Republic, knocking into the giant cooling towers. At Stone Oil, a twenty-five-hundred-barrel tank that lived through the first onslaught is now lit up, and its domed roof has been launched like a saucer 350 feet away.

Because there are so few people left alive on the immediate waterfront, because so many people in the city have evacuated and fled onto the highways toward Houston and Galveston, only two men are reported killed in the *High Flyer* blast.

One of them is identified as a student mortician, someone sent from Galveston and hit by falling shrapnel.

The identity of the other person, apparently standing closest to the docks and sliced in half by falling steel, is a mystery.

At his fallen-down house, Ceary Johnson has lost his frame of reference. He looks outside, in the darkness of early morning, and he can see Lucifer's fireworks. Thousands of those bright red pieces are descending from the sky-high column of flame. By dawn, he is still convinced that it is Judgment Day. Things like the eternal ledger are being settled. His neighbor in The Bottom, Rev. F. M. Johnson, has no explanations all morning as he braces himself against any more explosions and looks out over the people falling on top of the bodies they have come to claim at the black-owned diner on the south end of Sixth Street.

As Thursday's sun comes up, the first extremely tentative death estimates are being compiled by IBM businessmen from Houston who have arrived with ledgers and adding machines. The *New York Times* has suggested in its front-page headline that twelve hundred are dead. For now, a hundred bodies have been positively identified, some of them plucked from the bay and horribly swollen by the chemicals and water. The flames and petrochemicals brewing inside Monsanto have blunted any detailed accounting of the four hundred or so employees who were in the plant at the time of the explosion.

Thursday morning there are also many more strangers moving through the city: investigators from the Coast Guard, the Bureau of Mines, insurance companies. There are additional FBI agents. Longshoreman Julio Luna is someone they want to talk to. He is the first person to have spotted smoke. He's somewhere in the Texas City area: people have seen him volunteering at the Red Cross shelters; people have seen him at Camp Wallace—a shuttered military base that has been reopened to house refugees and more dead bodies.

The rumors are growing exponentially:

That second person killed in the High Flyer *explosion was Father Bill Roach.*

He was the only one left on the destroyed docks, standing utterly alone, watching the failed, straining tugboats retreating from the heaving *High Flyer* and pushing into the yellow night smog on Galveston Bay. There was no one else on the fiery waterfront by then. Anyone sane had left. There was a person that the dazed tugboat sailors saw back onshore, walking among the bodies that hadn't been collected, kicking over the rubble and digging into twisted metal piles and pulling dying people out and offering them the last rites.

It was Roach kneeling dockside when the High Flyer *blew*.

And on the cots at the Red Cross shelter, people are slandering the French crew:

They threw goddamn cigarettes into the hold. They refused to run a hose into the holds. They were worried about ruining their cargo. They were sliding down the moorings before the explosion. They were holding on to their damned luggage and yelling for cabs to take them downtown.

THURSDAY MORNING, Elizabeth Dalehite is determinedly driving back through the smoke and fumes to downtown Texas City. As per the regulations laid down by Curtis Trahan's Dead Body Commission, she and her son will be allowed to pass through the police barricades. Coming back into the city center, they see there are still blazes all over the waterfront.

The twelve fire departments that have descended on the city have run out of water. Pumping it out of the bay, now a sea of explosive chemicals, is dicey. The water is thick with jagged hunks of metal, charred timbers, body parts, and clumpy circles of dead fish. An aluminum tank of isopropyl acetate is releasing its contents, adding to the endless gallons of acids and petrochemicals gushing from the Monsanto plant, Union Carbide, and the riddled railroad tank cars. The chemical ponds are like bubbling drowning pools, and the ground is saturated, incapable of absorbing any more.

Trahan sends runners to alert Hill and everyone else on the city disaster committee that there is a 10:00 A.M. emergency meeting.

He has already sat in on a briefing in the City Hall war room where the assembled fire chiefs have agreed that the city has untold petrochemical blazes. It might take until the weekend before the fires, constantly feeding themselves on the propane and chemicals, begin to wither.

AS ELIZABETH DALEHITE reenters the lower half of downtown Texas City, there are powerful benzol tank eruptions on the grounds of Monsanto. And there are the long lines of hearses shuttling back and forth from McGar's gas station—the place where the horrible essence of the Dead Body Commission is on grim display. The biggest need in Texas City is for volunteers to staff the place that has become the city's makeshift morgue.

The first bodies brought inside are those of ace WWII pilot Johnny Norris and the passenger who had paid him $10 to take a spin over the funny smoke.

Inside the L-shaped building, about eighty feet long and forty-five feet wide, there is an impossible smell of formaldehyde, lime powder, smoke, and death. There is the acrid tang of kerosene, bathtubs of it, used to soak off the oil and molasses from the victims. The floors have been lined with sandpaper so that the volunteer embalmers won't slip in the pools of blood. The army has furnished cots, but there aren't enough. Worktables, even grease racks, are being used as embalming tables.

High school student Charles Cozzens has convinced his mother that he should go to McGar's and sign up to work alongside the 150 embalmers who have come from across the country:

"Mother was reluctant to let me but said if I thought I was up to the job, to go ahead.

"They painted my cuticles with an iodine compound, placed long rubber gloves on my hands, put a surgical mask on my face, and applied a camphor-scented liquid to the mask to make the work more tolerable. The shop was equipped with steel vats, about three by three by six and a half feet, filled with formaldehyde. The remains were disinfected in the vats and taken to the gym for identification.

"While there, I saw remains brought in tied up in army blankets. . . . A complete body was in one of the vats . . . apparently blown out into the bay."

The husband of fourth-grade teacher Beryl Wages has also volunteered to work in McGar's: *". . . cleaning bodies, first with gasoline to remove the oil and grease, and then soap and water. The bodies were transported in baskets and all did not contain the same number of parts. Some had two arms and two legs and some had one or three arms or legs. Some were women or children. It was a heartbreaking job."*

As the untrained residents of Texas City process the remains of their neighbors, friends, relatives, and classmates, some simply remove almost all the identifying shreds of clothing, watches, rings, earrings, tie clasps, and belts. Hundreds of bodies, whole and partial, are moving through McGar's and then on wheelbarrows, stretchers, or carts across Sixth Street and to the floor of the high school gymnasium. Because of the earnest efforts to cleanse the remains, many valuable bits of identification are removed.

After the bodies and remains are delivered to the gym, they are covered with brown Fourth Army blankets, or gray navy blankets, and arranged in neat, orderly rows under the basketball hoops. Volunteers loop paper tags with metal wires around a toe, a finger, a wrist, or, if necessary, a corner of the wool blanket. In the mad scramble to process the bodies, hundreds of ID tags are needed: tags that will hopefully have a full name, a possible positive identification, written on them. Some tags will have only some hint of information about the deceased, such as mention of a tattoo or a birthmark. Some tags will bear no information about the body under the blanket, and those will simply be marked with a ledger number.

There are no more ID tags to be found anywhere in Texas City, until someone from one of the area police departments has an idea. And now, each of the more than five hundred bodies that are moving through McGar's and then to the high school gym has a police traffic control tag on it.

On the blank side, the volunteers have hastily scrawled whatever information they have gathered: a name, if that positive identification has been made; where the body was found; old scars; approximate age; gender.

When the body tags are flipped over, there are words printed in bold type on the other side: YOU HAVE VIOLATED A TRAFFIC REGULATION.

AT THE GYMNASIUM on Thursday, it is the wicked rising tide, the pitiful convergence, that is binding Texas City. This is where the people of Texas City, the ones who haven't met one another, are joined. Elizabeth Dalehite is en route with her son. Forrest Walker will be there with his mother. Christine Baumgartner will be there with her children. Kathryne Stewart is coming. Rev. F. M. Johnson is being asked to walk through the gym along with members of First Baptist who can't bear it alone. The Mikeskas and employees of The Company will arrive. Ceary Johnson knows he has to go there and walk with the widows of the men on his longshoremen gang. And there are also dozens of the faithful who have not heard, or who refuse to believe, that Bill Roach died at a hospital in Galveston.

Outside the gym, the chalkboards salvaged from Danforth Elementary School have been set up, the spelling lessons on them erased. All day, victims' names are added to the boards as more identifications are made. Silently, families mouth the names; when they spot a familiar one and see the number next to it, they know where to go in the gym. Each of the corpses, identified or not, whole or not, has been assigned a number.

Inside, there is a slow procession under the dull light coming in through the cracked, high gymnasium windows. People are clutching flashlights or candles because the smoke is still in the air, still wafting in from the endless fires. Quicklime has been scattered about, but the smell, instantly familiar to some World War II veterans, is still overpowering. Mostly, people move in silence among the aisles. Throughout the day there are sharp noises echoing in the gym—the thunderclap of an oil tank blowing out, a backfire from one of the Fourth Army trucks. When the sounds hit the gym, the searching families reflexively fall to the floor, huddling alongside the hundreds of corpses.

The bleacher seats have been unfolded, and each row is filled with little clusters of personal effects that had not been discarded by the volunteers. There are money clips, hair clips, watches with smashed faces. Each of the clusters has been assigned a number to correspond with the number written on those police traffic control tags.

All morning, Kathryne Stewart's relatives have been trying to talk her out of going to the gym. She finally says: *"I must go."* At the gym, she is led by the arm, but she pulls away. *"You don't need to help me, I can make it."*

As she moves down one of the aisles of bodies, she stops alongside a blanket marked Number 158. The tag has tentatively identified the victim as someone named "Premore." It is a tall man, his foot twisted completely around and sticking out from the blanket. The face, hit by something blunt directly between the eyes, is completely bandaged. Strangely, the hairs on the chest are not singed. Almost in a matter-of-fact way, Kathryne says: *"Why there's Basil, right over there."*

Of course, she would recognize her husband's body.

"To me, he was perfect. You know, I've been married to him for ten years, and I have never known him to be unjust. I've never heard him raise his voice in unkindness to any person. And no one could even trap him into saying critical or nasty things about anybody. . . . He was brilliant and he was handsome."

Meanwhile, with his mother, Forrest Walker walks into his high school gym, into the familiar place. He and the other seniors have been preparing for the upcoming prom, an event that was going to be held inside the gym.

"Some were so charred that they looked like large slabs of burned bacon. Even though quicklime had been thrown on them, the stench was so bad that one could hardly breathe without gagging. . . . Rescue workers brought portions of bodies wrapped in army blankets."

He and his mother search through the bodies.

There is no word from Monsanto officials about his father's whereabouts. The Monsanto president, Edgar Queeny, is supposedly en route to Texas City. Finally, up on the bleachers, Forrest spots his father's pocket watch and his keys. They are marked Number 244.

Forrest retreats into the rows of bodies and finds Number 244. It is a small bundle, wrapped in a blanket, that measures three feet long and two feet wide. In his high school, under the drooping basketball nets, Forrest feels himself absorbing something that he knows will never go away. His mother seemed more angry than sad. She thought of herself as a good Christian and questioned why things could end this way. He knew his mother would carry on. In time, his mother would marry someone whom his father had known.

"When you are young, you think that life will last forever, but in that high school classroom, I learned otherwise. Strangely, even after the grisly discovery in the gym, the remnants of my hopes for finding him safe lingered on for several years in a recurring dream where we found him alive and well. Death of a loved one is always difficult, but at the age of seventeen, it simply felt unreal. Yet what was once impalpable has now become part of life. . . ."

Members of the Mikeska family, along with employees of The Company, return to the gym regularly, hoping for any news, any newly arrived body that might bear the faintest similarity to the man who ran the waterfront.

It seems that half the surviving citizens of Texas City are there as well, looking for Henry Baumgartner.

By day's end, there are no reports about either man.

They are still vanished.

By the end of the day, almost half of the two hundred bodies in the gym will remain beyond positive identification. The hopeless families wind their way to their last stop in the high school. The FBI and the Houston police department are overseeing the fingerprint desk.

Numbed parents and children are handing over military discharge papers, immigration papers, work IDs, and even a silver cigarette case that might have fingerprints. For hours, Houston police officers wave magnifying glasses over anything that has a fingerprint on it—and compare it to the barely usable fingerprints from the horribly burned bodies.

Finally, Elizabeth Dalehite and Henry Dalehite Jr. enter the gym, looking for the sea captain. She has always been resolute, as strong, in her way, as her husband. She comes from a family of ten children; she is the third youngest. She never wanted to live anywhere else but close to the Gulf of Mexico, about as close as you can get. As she and her son walk through the gym, her body is racked with pain. Her arms, legs, chest, and scalp are filled with dozens, maybe hundreds, of pieces of glass. They will come out of her body for several years.

There is no sign of her husband.

She and Henry Jr. go back home and wait.

Later in the day, the phone rings. A friend named Clark Thompson, a budding politician married into a prominent Galveston family and someone who knows everybody, is on the line. Thompson speaks carefully, as gently as he can. He has gone to the high school gym by himself to see if he can identify anyone. He has news. He has seen Henry in a row of newly arrived bodies.

His tag indicates that he is Number 122.

Henry had been sliced in much the same way that Elizabeth's statue of Mary had been sliced. A piece of a ship cut him the same way that her favorite holy statue had been cut.

Elizabeth absorbs the news, and then the sea captain's house on Offat's Bayou is filled with screams. In the old home, Henry Jr. listens to his mother finally breaking down.

He knows that his father, one of the legends of the Texas coast, will have to be buried in the family plot, just alongside his father's infant daughter.

AT THE MORNING briefing at the White House, the disaster in Texas City is the first order of business.

White House press secretary Charles G. Rose says that President Truman has been in touch with federal agencies, with General Dwight Eisenhower, with other officials, and that the U.S. government is prepared to extend all aid to Texas City. Rose says that President Truman has no specific steps to announce at this time because it appears that the relief operation under the direction of the Fourth Army and the Red Cross is already taking its course.

In Texas City, a Western Union delivery boy has jumped up the steps to City Hall with a telegram from President Truman to Curtis Trahan, and now the small-town mayor is holding it in his hand. It is one of many that he is receiving. General Eisenhower has sent one as well:

"As Chief of Staff of the Army, I assure you that we are willing and anxious to give any aid within our power and resources. Do not hesitate to call on us."

There are messages from the French ambassador to the United States; flags in France are being flown at half-mast. Canadian government officers have wired in condolences and offers of aid. There are telegrams from British officials in London:

"Londoners who bore the strain of wartime blitzes appreciate the immensity of suffering and associate themselves . . . in offering the deepest sympathy to those bereaved."

And Curtis is given a message that he will be getting a copy of a resolution that Senator W. Lee O'Daniel is speeding onto the floor of the

Senate. O'Daniel thinks the ammonium nitrate explosion might very well have been an extraordinary act of sabotage by either domestic or international terrorists. In the last several weeks there have been a string of smaller but still horrific disasters across the country—mine explosions, train wrecks, uncontrollable fires, and even massive eruptions inside factories. Now O'Daniel is publicly demanding an immediate, full-scale investigation by the FBI—by any government agency that can get to the bottom of the disaster and root out any terrorists who might have used ammonium nitrate to level a city and disrupt the flow of oil and chemicals across the nation. The stakes were high in the new Cold War, and O'Daniel and others in Washington were convinced that the new war was going to be unlike anything the United States had ever faced—it was a war that would involve attacks on the homeland, attacks directed at America's most symbolic and vital centers of commerce.

O'Daniel's resolution says:

> "I realize that this series of explosions could have been accidental . . . but in view of the high tension on international affairs and due to a rather large number of fires, explosions, railroad wrecks, and other disastrous occurrences in this nation lately—all so nearly resembling disastrous occurrences which preceded our entry into the last war, due largely to communistic underground activity—I believe it is the duty of this Senate to conduct a full and complete investigation into this Texas City disaster and to start the investigation immediately."

In his office, the small-town mayor weighs the flood of inquiries and messages from the White House, the FBI, General Eisenhower, and various overseas leaders. Last night, Curtis had wondered, again, if his city was dying and would simply disappear. He feels the same way now.

When the *High Flyer* went up, Trahan and Hill had been in the City Room and were thrown to the floor. The men were bracing themselves, and the dark room was completely illuminated by the sky-scraping flames. When the rumble finally faded away, Curtis crawled back to his feet and tried to find out what happened to Edna and the children.

She had been at home, singing to the boys. They had refused to sleep in their beds, and she had made pallets in the living room. She sang some more lullabies, and then she tried to listen to the radio. With the boys sleeping, Edna walked through the house, opening the windows that weren't already broken. The house had rocked, again, and then Curtis had called. They knew, at least, that everyone in their family was alive. She assumed he was going to be okay. And she felt that while she knew Curtis so very well, she couldn't remotely understand how he was contending with everything he was experiencing.

He has been gone all night, and now it is ten A.M. He has to get his emergency meeting under way at the City Hall battle station. Brigadier General J. R. Sheetz from the Fourth Army is there. So is John Hill. The FBI. Officers with the 32nd Medical Battalion. The dozen other emergency committee members.

Curtis informs everyone that there are even more hastily delivered telegrams coming from the White House, cabinet officials, and other leaders in Washington, D.C.

Secretary of the Navy James Forrestal, also the nation's first Secretary of Defense, has wired Trahan and told him that the Commander in Chief of the Atlantic Fleet has put the hospital ship *Consolation* at Curtis's disposal. The surgeon general of the U.S. Army has been ordered to fly to Texas. Several Pentagon–War Department officers who normally coordinate America's chemical warfare programs are also on their way. The army is flying in its ranking medical examiners, along with surplus military embalming fluid and fingerprinting kits.

Then Trahan announces that he has talked with the governor of Texas and they have reached a decision to designate the American Red Cross as the lead disaster relief agency. The Fourth Army generals agree to offer military planes, vehicles, and soldiers to serve under and alongside the Red Cross coordinators arriving from New York, Virginia, Georgia, and Missouri—and from the site of that Force 5 tornado that had engulfed large parts of the Texas Panhandle and southern Oklahoma just last week.

The army and the Red Cross are beginning to ship the first of 2 billion units of penicillin and 5 million units of gas gangrene antitoxins and tetanus antitoxins. The penicillin shipments, from Red Cross medical warehouses in Brooklyn, Chicago, and St. Louis, will almost deplete Red Cross emergency supplies—and will threaten to erase the nation's readily available supply of the new wonder drug. Army transport trucks are already moving four thousand units of blood plasma and nine hundred units of whole blood from Galveston and Houston. Thousands of barrels of distilled water are being dispatched from underground civil defense shelters. Five hundred more gas masks are to be delivered from Fort Sam Houston, along with ten thousand blankets and twenty-six hundred cots. Three hundred gas masks are being flown from the naval air station in Dallas, and additional gas mask requisitions are being sent to Fort Crockett.

Curtis knows that the first navy hospital plane, flown north from Corpus Christi, landed yesterday at 10:30 A.M. Now Curtis is told that hundreds of other military and commercial planes are landing in Houston, Galveston, and Texas City. Squadrons of C-47s from Brooks Air Force Base in San Antonio arrived by noon on Wednesday.

There will be four thousand Texas City–related relief flights made from as far as California, Georgia, Massachusetts, Illinois, North Dakota, Colorado, and Pennsylvania. Arriving in force on Thursday, an estimated two thousand doctors and nurses and medical school students, either native to Texas or flown in from around the country and overseas, are going to be involved in the medical relief effort. Ten school nurses even arrive from Dallas, and they are given the job of overseeing "the milk station"—finding food for infants.

In their records, medical personnel dryly note that there are some victims, up to five miles away, who have "received injuries from flying objects."

NATIONAL GUARD UNITS are stringing up emergency light poles. Local Boy Scout troops, white and Negro, stand at the bus station and the Texas City airport to serve as guides for incoming relief workers. Refugee shelters for two thousand white families are already open in Houston, Galveston, San Antonio, and other towns in Texas. At Camp

Wallace, just reopened for the emergency, 1,250 people are being sheltered the first day.

Western Union operators begin to receive messages from Mexico, Canada, and Europe—primarily from people wanting to learn the names of the injured or deceased. A total of twenty-seven thousand inquiries are patched through by telegram, phone, and shortwave radio.

Amateur radio operators call cities for emergency rations of Type O blood; the message is picked up in San Francisco, and twenty-four pints are put on a plane. Trahan's office fields an offer of help from IBM—the company would be willing to assign more statisticians to compile casualty reports and to file and catalogue all of the twenty-seven thousand inquiries. Braniff International Airways says its corporate flagship DC-3 can be used for news crews. Eastern Airlines, Pan Am, and American Airlines provide free transportation for any media, doctors, and Red Cross officials coming to Texas City. From Kansas City, TWA outfits one plane for doctors, nurses, and medical supplies.

Negro refugee families, some three hundred of them, have been steered to a segregated camp set up in the "colored" school in the small Texas town of LaMarque.

Curtis understands that irony.

LaMarque was the other Gulf Coast city that been secretly plotting to annex the companies on the waterfront before Texas City could get to them.

When Curtis had heard that LaMarque was going to file the necessary paperwork, he knew that the entire future of Texas City was at stake.

If there was ever any hope of finding the money to rescue El Barrio and The Bottom, he had to move fast to annex the companies on the waterfront—and he had to move fast to beat LaMarque.

Now, of course, most of El Barrio and The Bottom was utterly destroyed.

The black residents of Texas City were being sent to that segregated school in LaMarque.

———

APRIL IS THE cruelest month.

It is almost biblical, the sweep and scope of the devastation across the United States in April 1947. There is so much devastation that W. Lee O'Daniel, the senator from Texas, can't help but think that some of it has to be part of a well-orchestrated plot by terrorists who have been quietly planted in the United States:

Eight men are killed in a mine explosion in Terre Haute, Indiana. Eighteen people are killed in a fireworks factory explosion in Clinton, Missouri. Sixteen people are injured when the Santa Fe Super Chief derails outside Raton, New Mexico. Nine are dead in the Exeter, Pennsylvania, mine explosion. Two are killed and thirty-five passengers are seriously injured in the Leverett, Illinois, train wreck. Forty are injured in the wreck of the Gotham Limited in Columbia City, Indiana. Eleven homes are destroyed in a devastating Cumberland, Maryland, fire. Families are evacuated from five burning houses in Cold Spring Harbor, New York. Fifteen families are left homeless after a raging apartment house fire in Antlers, Oklahoma. Three homes are destroyed by a fire in Alsea, Oregon. Twenty-nine families are homeless after spreading fires in Baltimore.

Then, of course, there were the natural disasters:

The Texas-Oklahoma tornado killed 185 people. Two people are killed and 2,398 homes are damaged in Michigan floods. Sixteen people are killed in a tornado that has pressed down on the Missouri counties of Crawford, Pemiscot, Phelps, and Worth. One is dead, twenty are injured, and 475 homes are damaged in two days of flooding in Pennsylvania. Six hundred homes are damaged by a pounding hailstorm in Goliad County, Texas. Twenty families are displaced by a rattling earthquake in San Bernardino, California. Two hundred families are evacuated in swiftly rising floods in Cook and Will Counties in Illinois. Twelve are hospitalized after a raging windstorm in Thibodaux, Louisiana.

Now, in Texas City, by the end of the day, it is impossible to know exactly how many people have died.

Three hundred eighty-five bodies will be processed through McGar's gas station and then through the Texas City High School gym. The one confirmed casualty from the *High Flyer* explosion—the student mortician who had volunteered to work in McGar's—is one of the last to be added.

When his body is brought to the gas station, workers lay him down on the exact same cleansing and embalming table that he had been using to process the dead in Texas City.

Curtis knows the story. He also knows about the reports of scavengers riffling the pockets of the "floaters"—the bodies swollen by tissue gas—in the bay. In his office, Curtis suspects that no one will ever understand exactly what happened in Texas City. He doesn't know if he will ever find out the exact truth about Bill Roach.

Since the twin explosions he has been home for only two brief visits, about a minute each, to check on everyone. Finally, at the end of the second day of what the national newspapers are already calling the Texas City Disaster, he remembers to order the flag outside City Hall to fly at half-mast. He has signed the order condemning the city's drinking water. Now it is dark, still almost impossible to take a deep breath because of the sudden surges of smoke and chemicals.

Driving home is grimly surreal. Someone has crossed the legs of a battered mannequin and put her on display in a crushed department store window. Soldiers, four and five at a time, roar by in Jeeps with half-dressed bodies dangling out the back. Curtis recognizes friends who have gone into their closets and wooden chests to take out their old military uniforms. Now, dressed in the same clothes they had worn during World War II, they're patrolling the streets, sweeping up glass on Sixth and stopping to carry someone literally on their shoulders.

Under the emergency lights strung up on the sidewalks, it's easy to see the mad jumbles of footprints in the endless patches of dried blood.

And, at St. Mary's Church, Bill Roach's church, the building looks as if it has surrendered to something inevitable and finally fallen in on itself.

Back in his once-familiar neighborhood on the edge of the city, the small-town mayor walks up to his house where the live oaks and the tallow tree are still, improbably, standing. Inside, Curtis hangs on to Edna and his two sons as if never to let them go. He is incapable of expressing everything he has seen and endured.

"There really are no words."

The City

April 18, 1947

As Curtis finally tries to find his first sleep, there is another explosion at 2:00 A.M. No one is even sure where it has gone off. It could have been an oil tank; it could have been something inside one of the melted warehouses. Calls are made, and the report is like all the others: the roar came from the waterfront. No one is injured. No one is down there.

With dawn, there is a grinding, grating noise rising from Bay Street and Third Avenue South. Dozens of Houston bulldozers and army trucks with winches are roaring back, trying to peel off the first towering layers of downed walls and the skeletons of dozens of cars.

Volunteers, National Guardsmen, and army regulars are working shoulder to shoulder running chains around sheets of metal and flagging the men working the winches and pulleys. For hours, they work to pull debris off any crushed victims, autos, and boxcars.

Families nervously wait on the edge of the activity, peering through the billows of fire and dust to see if anyone is coming out alive.

The tanks at Humble Oil and Republic are still ablaze. The Monsanto plant is alive with tall tongues of flame inside boiling pools of benzol. The winds slap at the bay water, but they do little to shove the hovering black blankets of smoke away from the waterfront.

Other families are pacing outside of McGar's and begging for information. Some have been reduced to chasing newspeople down the street and asking them if they have any word on their parents, husbands, wives, children. Alicia Garcia, who lives down the block from Florencio Jasso, tells a crowd of reporters that she thinks three of her babies are dead—her three-year-old son, Joe, her nine-year-old son, Vincent, and her thirteen-year-old son, Troy. And when the photographers raise their cameras and the flashbulbs pop, other families instinctively fall to the pavement and hide their faces.

The photographers stare for a while until they realize that any sudden light, any sudden flash, any sudden noise are horrific triggers for almost anyone who was in Texas City during the explosions.

At City Hall, John Hill is growing impatient with the media. He has been doing most of the talking to the news crews. He has gotten familiar with the routine, the way he is instantly swarmed when he steps out of Curtis's bunker in City Hall. He has decided to hold a daily press conference at 10:00 A.M.

The reporters are hollering, wanting to know if there is going to be another explosion. On Friday morning he tells them there shouldn't be any more mammoth explosions, but he is still recommending that anyone who had escaped Texas City stay away until the weekend. Even with the fires still gripping the waterfront, Hill optimistically adds that there won't be another detonation like the ones from the *Grandcamp* and the *High Flyer:*

"*We're in good shape now.*"

Curtis steps out, too, thinking that a statement from the mayor might mean something:

"*There is no imminent danger of further explosions.*"

The pool of reporters, from the national war correspondents flown in by the *New York Times* to the local guys from Galveston, are not satisfied. They are pressing hard, demanding that they be granted permission to go to the waterfront. Hill, operating as deputy mayor of Texas City, issues an order:

"*Any newsman, wire service man, radio announcer, or photographer found in the restricted area will be immediately arrested and made to leave the city.*"

The reporters bicker and complain. Roy Hanna, a frustrated editor at the *Galveston News*, decides to beat Hill. He wants to fake his way past the barrier set up by the Texas Rangers, state highway patrolmen, Texas City police, and the Fourth Army soldiers. Hanna breaches the security line by driving up in a Jeep with a crushed roof and saying that he is part of a rescue team sent to carry out the remaining bodies on the waterfront.

As Hanna picks his way past the wreckage in an open field west of the refineries, he passes one of the smashed Piper Cubs that was snatched out of the sky. He moves on. There is a dead mule with a giant piece of steel protruding from its neck. Walking closer to the smoldering docks and through the small neighborhoods alongside the Monsanto plant, he is joined by three geese waddling out from underneath a flattened car. A half-dead dog is bleeding, hiding and howling inside its battered doghouse.

Hanna crosses the road alongside Monsanto, and he is losing track of the number of smoking, burned-out cars. He thinks there are hundreds. On his left one of the giant Humble Oil tanks is spitting fire, and Hanna ducks the waves of thick, greasy smoke rolling back and toward the city. Hanna walks alone along the concrete slip that looks as if a giant hammered at it. He leans over, and he can see that the entire slip has been chopped in half. As he leans a little more, he is suddenly inches away from the remains of a fire-blackened face. Hanna, who didn't know what he would see when he talked his way through the barricades, is screaming for a stretcher. Three rescue workers crawl over the mountainous piles of ruined metal and tangled pipes.

The workers pull away mounds of charred wood and concrete and reveal the rest of the corpse. It looks like a man wearing greenish cowboy boots. Nearby, in the water, there is another oil-smeared body, bobbing facedown amid the wood, jagged metal, and electrical wires. The rescue workers ignore that corpse.

Hanna moves on. There are bodies everywhere. A rescue worker, his face blackened by smoke, is up ahead and yelling for another stretcher. He has found less than half of what appears to a female office worker. Hanna walks a few more yards south, trying to make his way through the craters, the smoking boxcars, the fallen buildings, and the melted warehouses. He wants to look at the main slip, where the *High Flyer* had been docked.

There are more distorted bodies splayed at odd angles on the makeshift path.

Hanna stops and runs back for someone to please bring another stretcher. He is shouting. A volunteer stumbles up with a stretcher, and they retreat to where Hanna had spotted a corpse. Hanna and the volunteer carry the body for thirty minutes, until they can find a car to take the corpse to McGar's garage.

As the car pulls away, a thought wells up in Hanna's mind:

"How many more bodies and pieces of bodies will be discovered after the fires have died completely? I don't know. Some will never be found."

EDGAR QUEENY, THE intimidating president of Monsanto, has shipped a news release to the Texas City newspaper. He wants it there before his plane lands on the ground in Texas and he is driven to Texas City.

By late yesterday, his public relations vice presidents realized that Monsanto and any other major corporation in Texas City would be haunted by the grim images dominating the front pages of every newspaper in the world, including the *New York Times.* There are 154 Monsanto employees dead, 200 Monsanto workers hospitalized. Anyone else in the building was injured in some way.

And yesterday, the *Times* ran the same kind of three-deck-high headline across the entire top half of the paper that it used during the attack on Pearl Harbor:

BLASTS AND FIRES WRECK TEXAS CITY OF 15,000; 300 TO 1,200 DEAD; THOUSANDS HURT, HOMELESS; WIDE COAST AREA ROCKED, DAMAGE IN MILLIONS

Now reporters are already beginning to compose stories questioning what role the "steel band" had in the explosions.

Was something inside Monsanto detonated? What was Union Carbide working on? Why were Monsanto officials shipped to Texas City from Oak Ridge, Tennessee—where part of America's nuclear arsenal was being nurtured?

The questions range to the obvious and the undeniable:

Why was the Monsanto plant built so close to the dock—and, more important, why was it built so close to hundreds of homes?

Queeny told his assembled executives he had decided that Monsanto would rebuild their entire plant. His technical director, who also is the president-elect of the American Chemical Society, says it can be done. Most of the plant had been refurbished at government expense during the war years anyway.

And, besides, Monsanto is already planning to file a claim with its insurance carrier, Lloyd's of London—a claim for the largest single-risk loss in the history of insurance. Queeny suspects the insurance company will pay out. Given the endless chronicling of the tragedy—and given the brash statement by the Monsanto president that the company would rebuild—he is presenting the insurers with a fait accompli:

The survivors in Texas City have been promised a new plant—and the insurers will have to pay out to Monsanto or risk a barrage of negative publicity. Queeny's bet will be right. In time, when it is eventually settled, Monsanto's insurers will be forced to pay what was then the largest sum ever paid for a single loss, estimated at almost $100 million.

Today's *Texas City Sun* looks like a flimsy handout.

Printed on handbills because the explosion crippled its normal printing operations, it has the simple announcement that Monsanto executives from St. Louis are saying that they will rebuild. There is also news that Monsanto executives will set up a temporary office in Texas City and immediately begin dispensing checks in the amount of $1,000 to the family of each Monsanto employee killed.

In Texas City, if there is resistance to the idea of Monsanto rebuilding its massive chemical plant, not a word is uttered publicly.

There is, instead, widespread relief. It is saluted as industry's instant belief in the future of Texas City. There will be jobs again. Someone, at least, thinks that the city is worth reclaiming. An entirely new chemical plant, one of the largest in America, will be built along the waterfront in Texas City. It will be built in essentially the same place—and as close to the homes and families and churches as before.

The one man, of course, who would have fought the idea of a

sprawling plant built alongside so many homes is William Francis Roach—the priest who was finally, completely, consumed by his own extraordinary visions.

THE *GRANDCAMP* WAS supposed to have shipped out Wednesday, and then it was to have stopped in Galveston before moving to the deep, open waters of the Gulf of Mexico.

Galveston had been the site of the worst natural disaster in American history, the Great Storm of 1900.

Now, a dozen miles away, Texas City had been the site of the worst industrial disaster in American history.

Today's *Galveston News* has an editorial:

"It may be only a trick of fate which saved Galveston from the disaster which wrecked Texas City. The ill-fated French ship Grandcamp *was due to be shifted into this port Wednesday night. We need but little imagination to visualize the effect the explosion would have had on Galveston ... Galveston, with its own experience of death-dealing disaster in the past ..."*

By now, the scope of that death dealing is becoming clear.

Federal and state building inspectors are moving through what is left of Texas City, beginning an initial inspection of fifteen hundred sites. The inspectors are carrying bundles of red, yellow, and green cards—they put red cards on buildings that are uninhabitable; they put yellow cards on ones that should be used only for limited access; they put green cards on buildings that seem stable. By the end of the day, they have already placed red "condemned" notices on 539 homes.

Auto insurance investigators are beginning to pool their numbers to put everything in some context. Six hundred cars have been completely destroyed and five hundred additional cars have been heavily damaged.

The national railways, which own The Company, have also begun their investigations—a staggering 362 freight cars, many of them loaded with chemicals, have been obliterated.

And now epidemiologists with the Fourth Army have suggested that something approximating a plague could develop. It is simply impossible to rescue all the rapidly decomposing victims. Some will not be found for weeks—if ever. Disease is bound to take root and spread. The explosion has unleashed armies of rats and snakes in the ditches around the waterfront. There are dead animals, schools of dead fish. Homes have been abandoned with food on the tables, food in pantries, food in iceboxes—and all of it is rotting, smashed, exposed. Wastewater pipes have been cracked. The mosquitoes have come in hordes.

Hill asks a local radioman for access to the airwaves. He announces that the entire city will be sprayed with DDT. Federal health workers, from the U.S. Health Commission, will be moving through the streets and into homes, schools, and churches. The first wave of workers will be using handheld sprayers. Larger pump sprayers and canisters of DDT are being shipped in next week.

DDT, dichlorodiphenyltrichloroethane, was first produced in Austrian chemical laboratories in 1873. Paul Müller, a Swiss chemist looking for ways to improve agricultural production—just like the German chemists who had experimented with ammonium nitrate—discovered DDT's insect-killing properties in 1939. Müller's chemical compounds would be produced in mass quantities during and after World War II to kill the body lice that could lead to typhus. His DDT would be used on battlefields and even liberated Nazi concentration camps—and the wonder chemical would be credited with saving millions of lives from malaria and other insect-borne diseases.

His extraordinary discovery, like Fritz Haber's, would also be rewarded with a Nobel Prize. Like Haber, his work would be lauded for saving humanity. The presentation speech for Müller's Nobel Prize was given by a member of the Nobel Academy, and it read, in part:

> *"Towards the end of the Second World War, typhus suddenly appeared anew. . . . In this situation, so critical for all of us, deliverance came. Unexpectedly, dramatically, practically out of the blue, DDT appeared as a deus ex machina. . . .*
>
> *"The story of DDT illustrates the often wondrous ways of science when a major discovery has been made. . . .*
>
> *"Dr. Paul Müller . . . your discovery . . . is of the greatest importance in the field of medicine."*

Twenty-five years after Texas City was heavily sprayed with DDT, the chemical was banned by the United States following extensive toxicological studies.

DDT was shown to lead to virulent cancers and tumors affecting the nervous system, the kidney, the liver, and the reproductive system. Federally funded studies showed that DDT was "very persistent in the environment, with a reported half-life of 2–15 years."

In time, some people would begin to call Texas City something else—"Toxic City."

BEFORE THE END of the day, Hill is furious.

Trahan can see it; so can anyone else at City Hall.

Walter Winchell, the most powerful media personality in the United States, had been reporting on his national radio show that Texas City is in danger of a third high-intensity explosion. This one would finally shove Texas City off the map. Winchell has a rabid, loyal audience for his shows and his newspaper columns—and millions of Americans hang on his words.

In Texas City, a local radio reporter had slipped past the police barricades, gone by the Republic Oil refinery, and smelled the vapors from a naphtha tank. He called Winchell, and suddenly a catastrophic forecast is being heard around the world.

Hill, who had gone from being the "tail-twister" at the Lions Club to redressing the commanding general of the Fourth Army, knows what to do. He asks one of the local phone operators who have been pressed back into work to call Winchell's network in New York. Amid the emergency calls and the thousands of desperate inquiries from friends and relatives, the operator patches through to the ABC studio in Manhattan. She hands the phone to Hill. He issues a demand:

"Winchell had it totally wrong and I want an immediate retraction."

There is no apology, no retraction.

Hill steps outside City Hall and tells the media to please set the record straight. And then he realizes, like Trahan, he has not slept since the night of the fifteenth. He has been awake for sixty hours. He walks to the U.S. Army field kitchen in the city auditorium—where

Bill Roach had first been taken after he arrived at the City Hall complex. The soldiers stare at him and wrap him in a field jacket. They steer him to a cot with a blanket and a pillow on the first floor of the Municipal Building.

That day, Hill also sees a pipe fitter he knows from the Carbide plant. Tears are in the pipe fitter's eyes. He tells Hill he has been working as a volunteer embalmer inside McGar's garage for forty straight hours. And, he says, they had to cut his cowboy boots off because his feet had become grotesquely swollen from standing so long.

Hill looks at his friend. The incongruity—a chemical engineer dealing with the Fourth Army, a pipe fitter embalming bodies—is absurd. There are other absurdities:

All damned day long, in the middle of every goddamn new tragedy, Hill has turned around to see someone stalking him: A deep-sea diver and his agent are begging Hill to throw them some work, to give them a contract to do underwater work fishing out bodies from the bay. Hill repeatedly tells them no. They follow him. They are there when he gets off his cot. They are there, walking toward him, when the reporters peel away to file their reports after another impromptu press conference.

Finally, the muscled ex-jock stares hard at the deep-sea diver: *"If you don't leave me alone, I'm going to have you arrested."*

The Gangster

APRIL 19, 1947

THE ALLEGED biggest gangster in Texas has always let the most famous celebrities in Hollywood have the run of his plush penthouse suite—with the wood-trimmed bar, the sweeping verandas, and the breathtaking view of the sun bobbing over the Gulf of Mexico. He even let two of the biggest stars in Hollywood get married in that suite atop the most famous hotel on the Texas coast. Now he wants to see if he can pull in some favors.

He draws up a list of the names he wants to come to Texas City:

Frank Sinatra. George Burns. Gracie Allen. Gene Autry. Jack Benny and his "man" Rochester. Phil Silvers. Bandleader Phil Harris, heard on radio stations across the country. Actress Alice Faye, Harris's wife and queen of the Twentieth Century–Fox musicals for almost a decade. A-list starlet Marjorie Reynolds—she was the one on the screen with Fred Astaire and Bing Crosby in the Christmas classic *Holiday Inn*. The Page Cavanaugh band with the singer Doris Day. Diana Lynn, a superstar with a long-time contract with Paramount and someone who will play opposite actor Ronald Reagan in a film called *Bedtime for Bonzo*.

And, if he can get them, he'll add in Mickey Rooney and Red Skelton.

He'll send the whole damn all-star cavalcade across the nation.

First to Galveston, of course. Then Houston. New Orleans. Maybe New York and Chicago. Maybe he'll bring the entire revue back to

Texas and wind the thing up with cocktails and bushels of fresh crabs. Maybe even a nice high-stakes poker game at the distant end of his "pleasure pier"—the long, narrow, velvety nightclub called The Balinese that's built straight out into the Gulf of Mexico on increasingly steep piers and that's decorated with a tropical mix of palm trees, seashells, and prints of dusky women in barely noticeable sarongs.

Across the west bay from Texas City, Sam Maceo and his brother lord over the illegal slot machines, roulette wheels, and blackjack tables on this stretch of the Gulf Coast. During Prohibition, they ran liquor up from Mexico, the same way that it was being run down to northern cities from Canada. At night, all manner of barges and steamers would appear offshore, somewhere out by the sandbars, and then they'd dump barrels. Rowboats would arrive and scoop the barrels out of the shallow, warm waters, and the alcohol thirst of the entire state of Texas would ultimately be satisfied. Sam Maceo is a big-time player, and his tastes are eclectic and dutiful—he is consulted and involved in just about everything in Galveston, from politics to the local magazine, from the restaurant game to the city waterfront.

It seems as if he's been trying for a hundred lifetimes, but Maceo still wants something approximating complete acceptance. Italian Americans in Texas are still called "Dagos" and "Spanish." No one calls it the Mafia, but along this part of the Gulf Coast, people assume that the Maceos are involved in organized crime. Hell, they assume the Maceos are the ones organizing all the crime in Texas. They assume Maceo is working closely with the nearby New Orleans Mafia kingpin Carlos Marcello—and with Frank Costello, Marcello's direct crime partner in New York City and the man who heads the entire Mafia Commission in the United States.

To counter the raps, Maceo has started up his own glossy Chamber of Commerce–type magazine. He installed himself as the publisher and hired some art directors and editors; the whole thing is upscale, handsome, and sometimes there's a little bit of skin to be found in the swimsuit photos. He's been building Las Vegas in Galveston, offering the slots and new hotels and nightclubs to go with the Balinese, and he's even hired the piano player away from the Hollywood Plaza.

Now he has an idea.

Maceo, from his penthouse, can still spot the fireballs rocketing out of Texas City. It's time to call in favors.

He wants to put together the largest traveling all-star ensemble of internationally recognized celebrities. Maceo, because he can, gets one of the available out-of-state phone lines. He reaches Phil Harris at his home in Encino, California. The curly-haired, mush-mouthed actor and musician, a mainstay for millions of radio listeners, says he'll start organizing it.

It's short notice.

Harris knows that he can easily get Jack Benny. Probably Burns & Allen. Probably Sinatra. What about Gene Autry? That should play big in Texas.

Harris was the one who married Alice Faye up in Maceo's penthouse suite. And now, Harris tells Maceo that they should schedule the thing before the end of the month. It'll be big. The biggest postwar gathering of the most famous American celebrities ever to appear in Texas, maybe in the United States. Nothing will be even close. Maceo, who thinks big and is accustomed to getting anything he wants, likes what he is hearing.

He'll call Curtis Trahan, get the small-town mayor to be on the Galveston tarmac when the international celebrities arrive.

It will make a good photograph.

A FEW BLOCKS away in Galveston, Rear Admiral Gordon Finlay is already overseeing the first official inquiry into the Texas City Disaster.

An emergency Coast Guard board of investigation has been assembled and is taking sworn testimony from longshoremen who had been on the *Grandcamp* work gangs. Longshoreman L. D. Boswell, who is in charge of Julio Luna's gang, is on the witness stand. Several times during his testimony, he mentions that the highest priority on the waterfront had been how to "prevent damage to the cargo":

> *I had seen personnel on the decks of the freighter smoking ciga-*
> *rettes the day before the blast. . . . I asked somebody on the French*
> *crew if they were allowed to smoke and they told me no. . . . One of*
> *my longshoremen reported that he had smelled something burning*

down in the hold. . . . Somebody told us not to pour too much water
because it would "damage the cargo." I don't know who that
was. . . . Somebody told us to close the hatch and my men moved up
on deck. . . . We got off there when we saw the hatch "raise up" after
the French crew covered the ventilator to the hold and turned on the
steam to put out the flames.

The hearings will continue for weeks.

There will be lingering, painful investigations—by the FBI, by
Monsanto, by military investigators, by the leading arson experts in the
world, by Nobel scientists who had also worked on America's secret
nuclear weapons programs. There will be hundreds of thousands of
pages of sworn testimony in depositions, federal district courtrooms,
federal appeals courts, international courts, and finally the Supreme
Court of the United States.

The investigations will stretch over three decades.

They will pit the quiet, extraordinary heroes of Texas City against the
entrenched government bureaucracy—and will force them to face the
monstrous possibility that their own elected officials had not just turned
their backs on them . . . the officials had left them inside an eternal
nightmare.

SINCE THE *GRANDCAMP* exploded on Wednesday, the Texas City
High School football field has been used as the staging area for the two
hundred firefighters who had come from the far-flung departments
around the country—including inspectors from as far away as Los
Angeles. Each morning, the firefighters assemble and listen to Fred
Dowdy—one of the few Texas City firemen left alive—as he reads the
latest reports that have been raced to him from behind the police barricades. Since Wednesday, the biggest problem facing the firemen is
where to get water. Pumping it from the petrochemical bay is beyond
dangerous. The usual dockside pumps and reservoirs have been detonated.

Now Trahan has already told Dowdy that the firemen should begin
to clear off the football field.

A mass memorial service is set to begin later in the day, just as the sun first begins to set. As the firemen leave, preparations are under way. Funeral home operators are moving across the field, dotting it with floral crosses.

By midafternoon, there are long lines of cars searching for parking spaces along the shell roads. The sky is still clotted with pungent smoke and people are breathing through wet handkerchiefs pressed to their mouths. From the haze comes row after row of flatbed trucks bearing flowers, bouquets and enormous arrangements.

There are at least fifteen hundred people leaving the skeletons of their homes, and they will participate in the first mixed-race religious gathering in the history of the city.

Rev. F. M. Johnson is there. Ceary Johnson is there. The Negro men didn't have to be told they were allowed to come. With the sun setting, they simply join the slow parade heading north on Sixth Street, moving down the middle of the road and being joined by hundreds and hundreds of others on their way to the football field.

As Curtis Trahan and the others walk uptown, the already dark sky begins to release pockets of fat raindrops. When the drops thud from the still hovering smoke and onto car windshields, they look black, greasy.

As the mourners continue north, they pass an eerie army moving south to the waterfront.

Specialized corps of body hunters with thick gloves, jumpsuits, and gas masks are just now being dispatched to the upper floors and offices in the Monsanto plant—where the fires have taken permanent root, where hundreds of Monsanto workers had been staring out windows Wednesday morning and marveling over the pretty smoke.

Under the direction of the Fourth Army, the Red Cross, Trahan, and Hill, the search process has now been codified. There are several distinct groups: the cutters who slice away debris blocking access to the Monsanto plant and the waterfront; road-building crews who can secure safe paths for men bearing stretchers and ambulances; first aid crews who can attend to anyone still alive; body hunters who will go where no one else will go.

As the mourners proceed through the streets, it becomes the largest mass parade in the history of Texas City. They move forward, past the federal workers sent by the U.S. Health Commission and who are walking in teams and hand-spraying even more DDT into doorways, churches, and homes.

As they move onto the football field, Curtis and the others bow their heads for the forty-five-minute memorial service. There are no segregated sections like the ones at the Showboat Theater or at Frank's Café. Four refinery workers from Amoco are singing a cappella gospel hymns, oblivious to the splashing rain. When they are silent, a college choir from Texas State Lutheran fills the void by singing "The Old Rugged Cross."

Curtis listens as the white Baptist minister, Roland Hood, begins: *"We sorrow not as those who have no hope. We have hope which reaches beyond this shadow. . . ."*

Black and white are standing side by side.

There will be more binding moments to come, more things and moments to fuse what's left of Texas City in haunting, prophetic ways.

Right now, with night creeping in from the expanse of the Gulf of Mexico, the awful mixture of grinding, groaning noises seems to magnify from the waterfront. People who are at the memorial service cannot bear to turn their heads and look to the southern sky. The black shadow from the waterfront seems hopeless and never ending.

Tomorrow will be the initial burials. There will be thirty-eight straight funeral processions to the Texas City cemetery eight miles away.

And then, for the next ten days, there will be one procession after another during every hour of light.

The Priest

April 24, 1947

The first memorial service for Bill Roach is held miles away at the towering cathedral in Galveston. The casket is open and as people approach it, it is clear that there are no visible marks on his body. His face is pulled into a smile.

There are fifty priests in attendance, including his twin brother, Johnny. The people from his St. Mary's parish are there, too, many of them in stained, bloody clothes—the only clothes that they have left. They have driven over the causeway, the smoke and fires to their left, and it is almost as if they are leaving one world and entering another. In Galveston, only a dozen miles southeast, there is a fresh salt tang in the air. The clouds over the Gulf of Mexico are milk white and hung high. A group of toddlers, each one holding hands with the children on either side, is walking to the beach.

Bishop Byrne, of course, leads the service.

He has begun crying, and not many people have seen that before. Byrne tries to fight his way through it. The tears fall as he stops and starts: *"The Lord giveth and the Lord taketh away. Today we should let our sermon be the words of Our Lord: 'Greater love than this no man hath than that he lay down his life for his friend.'"*

From the cathedral, Byrne and the others watch as Johnny Roach wheels his brother's casket away. He is taking Bill Roach home, away from Texas City, back to Pennsylvania.

———

THE SECOND MEMORIAL service for Roach is on Sunday, in the Texas City High School auditorium. St. Mary's Church is still condemned by the building inspectors. Walter Winchell is still on the air, predicting another explosion, and there is a whole new wave of cars lurching out of the city.

Rev. Thomas Carney leads the service at the high school. As he rises to speak, that constant drone is heard through the shattered windows and broken doors. It is the jumbled roar of fire, machinery, sirens, and shouts. The faint but unmistakable smell of death and formaldehyde seems to have taken root in the floors and walls.

"Before Father Roach came here, I told you that a saint was coming into your midst. A saint came, and a saint has been taken away."

Carney leans into it and stares at the dismal people assembled in the auditorium. They need to know Roach's legacy, if they didn't already. They need, now, to appreciate the work he had been doing. Carney isn't the only one who knew about Roach's overly vivid, excitable prophecies. But now Carney is convinced that, before anyone else, Roach had seen the only real way for all of Texas City to converge:

"Father Roach was very close to you. I think he died just as he would have wanted to die. In this disaster, we have seen men of different political beliefs, of different religious beliefs, stand shoulder to shoulder and smilingly face death."

THE THIRD MEMORIAL service for Bill Roach is held, finally, outside his own church on Thursday morning, April 24.

The building is still in danger of having the roof beams cascade down, but the safety inspectors have agreed to let people mingle on the church grounds. Curtis had to go on the radio again and talk about the fact that blasts of orange smoke were bouncing over what remained of the waterfront. Bags of ammonium nitrate that hadn't been loaded onto the freighters were catching fire in the dregs of The Company's warehouses.

It was the same beautiful smoke. Curtis spoke calmly:

There is no danger. I promise I'll tell everyone in Texas City if there is any danger.

Curtis has even gone further.

He is urging people to move back to Texas City, to begin rebuilding their homes and businesses. The majority of bodies have been processed through McGar's. The high school gymnasium is almost ready to be returned to the graduating seniors, including Forrest Walker, the teenager who had discovered Curtis's son. And at Clark's Department Store, where Bill Roach used to watch in amazement as the women from Tillie's whorehouse shopped for frilly things, the owners have reopened under a Quonset hut.

Now a large stage-like altar has been built on the south side of St. Mary's. There are Mexicans and Negroes in attendance—including people from St. Elizabeth's Hospital for Negroes in Houston, the medical facility that the Roaches were willing into existence the exact week of the explosion.

By 7:00 A.M., the lawn is crowded.

By 8:00 A.M., people are clumped on the sidewalks, standing yards away. There are five hundred people now, some of them standing on pieces of rubble from the knocked-about building. Elderly women are in the shadows alongside the Spanish-style church. The sky is back to its usual concrete color. The humidity is coming on, and the clouds could burst at any second.

Father Carney, Roach's friend, has been invited back to deliver another eulogy.

There is a tall man in the crowd in front of him, somebody who stands taller than anyone else. It is Curtis, and he is wearing a suit and holding his hat. Curtis isn't Catholic, though some of his relatives are. This is his last chance to say good-bye to Bill Roach. Trahan waits for Carney to begin.

Carney talks, for a few seconds, about the fact that Roach's car was found near the heart of the blast. Some people have seen it in the photographs—it had been flung into the air and dropped upside down until it came to rest at the peak of a mountain of other burned-out cars.

Then, Carney says he knows what Bill Roach wants.

Curtis raises his head. He is going to fly out of Houston in three days. He is going to testify before a House Appropriations Committee and ask them to consider giving $15 million to Texas City to build its first hospital, to repair the drinking-water lines, and to relocate hundreds of homes away from the dangerous waterfront.

Now, Carney is staring directly at him and aiming his comments straight at Curtis:

"Mayor Trahan, Father Roach would want me to tell you to rebuild your city of God. Father Roach would want me to tell you not to be discouraged, but to rebuild your city into a greater and safer place in which to live.

"Father Roach was the most holy man I ever knew. He wants you to carry on, to reconstruct a better, bigger, safer city like the one he had envisioned."

CURTIS TRAHAN, the respected and well-liked World War II hero and mayor, suddenly found himself tested in ways he'd never imagined.
(Courtesy Houston Metropolitan Research Center, Houston Public Library)

SEVERAL SURVIVORS remembered how, in the weeks before the Texas City disaster, Father Bill Roach had talked about "blood flowing in the streets." Father Bill was loved and hated along the Gulf Coast for fighting racism, poverty, and injustice.
(Archives, Diocese of Galveston-Houston)

The *Grandcamp*, docked in Houston, a few days before
its fateful voyage to Texas City. *(Copyright © Houston Chronicle)*

FIREFIGHTERS, led by Chief Henry Baumgartner, immediately converged on the scene. All of the firefighters who responded to the fire on the *Grandcamp* would be killed in the initial blast; their remains would never be identified. *(Courtesy Houston Metropolitan Research Center, Houston Public Library)*

As explosions rocked the city every hour, some residents risked death to rescue their worldly possessions. In the background, flames and smoke are moving toward the blown-out remains of Our Lady of the Snows Church. *(Rosenberg Library, Galveston)*

Rescue workers toiled for months, searching for victims and the possible survivor. Almost every family in Texas City suffered a death, injury, or financial devastation. *(Courtesy Houston Metropolitan Research Center, Houston Public Library)*

THE TEXAS CITY disaster dominated world headlines, and
news organizations dispatched seasoned war correspondents
to cover the explosions. *(The New York Times)*

A DEVASTATED CITY.

THOUSANDS of homes, churches, schools, shops, and offices were damaged all over the city—and in other cities as far as 150 miles away. As people searched the rubble for survivors, townspeople found themselves brought closer together by tragedy.
(Courtesy Houston Metropolitan Research Center, Houston Public Library)

THE REMAINS of the *Wilson B. Keene* as it struggles to stay afloat after the *Grandcamp* exploded.
(Courtesy Houston Metropolitan Research Center, Houston Public Library)

THE BIGGEST names in Hollywood and the entertainment industry staged fund-raising events around the country. Frank Sinatra *(left)* meets with Curtis Trahan *(center)* and military officials before one of Sinatra's events in Texas.
(Copyright © Houston Chronicle)

SOME SAID only an "act of God" could bring a divided Texas City together. Here, the members of the different denominations gather to bury their unknown dead.
(Courtesy Houston Metropolitan Research Center, Houston Public Library)

The Unknown Dead

JUNE 22, 1947

THE LAST body to be officially removed from the waterfront is found, buried under debris, twenty-six days after the *Grandcamp* disintegrated.

It is taken, along with the remaining unidentified bodies, to the refrigerated mess kitchens at the former army base at Camp Wallace. McGar's service station–morgue has finally been vacated. The Dead Body Commission has told Curtis that the final death total will never be known . . . the brutal reality is that dozens and dozens of people were injured too badly ever to be recognized or found.

There are 405 identified dead. There are 113 known to be missing. There are 63 bodies unidentified.

For uniformity, the "official" total number of dead is now placed at 581.

Investigators tell Curtis they presume the final death figure is much higher—life in El Barrio and The Bottom had been so tenuous to begin with that death counts from those zones are more than difficult to obtain. No accurate census counts had been taken there, and now the homes have been so obliterated that whole families are disassembled. Meanwhile, there were an untold number of visitors in Texas City at the time of the explosions—relatives, friends, truck drivers, sailors, travelers, itinerant longshoremen, and merchant seamen.

There is no definite way of knowing who lived and who died—or who will yet die from complications.

Death certificates and medical records will be sealed, and no accurate account will ever be available to fix the number of children who died—though an internal Red Cross investigation shows that twenty-three of the missing are under the age of sixteen. Three of those children are under the age of six. At least nine children from Booker T. Washington School have perished.

Finally, there is the matter of the souls who have simply discovered the perfect, extraordinary moment to leave behind their jobs, their families, and everything else that had once been familiar.

Ruth Kempner, a Red Cross worker, echoes what many in Texas City suspect:

"People took advantage of the general confusion just to disappear. Some people were just never heard of. Usually there was enough left of a body to identify it, but it was an incredible opportunity, if your domestic responsibilities were too much for you, to disappear."

CURTIS'S CLOSE FRIEND Frank Doremus is the young pastor of the Episcopalian church. Curtis is an elder in the church—he holds the title "senior warden." Doremus always thought Curtis was steady. Doremus, who had been in Texas City for only a few months, always felt he could trust Curtis: *"A good man, a capable guy. He was loyal to the community, loyal to the church, a thoughtful man."*

During the first days of the crisis, when he finally found a free minute, Curtis went to Doremus. He did some chitchatting, but mostly Curtis was quiet. Doremus knew Curtis well enough to figure that he was asking for help, some sort of direction. Doremus bowed his head, drew close to his friend, and tried to talk Curtis through it.

Now, weeks later, Curtis is asking Doremus to perform the most visible act of closure. He asks Doremus to organize and lead a memorial service for the unidentified dead. There no longer is any hope of identifying the sixty-three corpses still in storage in Camp Wallace.

The nameless dead—black, white, Hispanic—must be laid to rest.

It has to be done before Texas City can even dream of moving on.

The city has moved quickly on many other fronts, though for years there will be a lingering, numbing, mixed smell of death, oil, and chemicals. Fishermen will refuse to lower nets anywhere within a half mile of the north slip. Children who had walked out on the levee with long cane poles for fishing now ride their bikes, instead, out along the entire length of the Texas City dike—going out to the end of the thing as if they are on a low-lying bridge that will take them into another distant part of Texas.

When the Second Army of the United States had finished occupying Texas City in 1915, the children grew accustomed to finding personal items left by the thousands of soldiers from Indiana, New York, Oregon, and elsewhere who had been steered toward the Gulf Coast. Now and for several more years, the children in Texas City will discover things in the amber waters of Galveston Bay, in the sodden earth along the levees and the collapsed docks . . . a cigarette lighter with the words CONGRATULATIONS FROM MONSANTO on it, a billfold with baby pictures and some coins inside, a rolling pin stamped with words in French and that once belonged in the galley of an oceangoing ship. Too, there will be many unsettling discoveries, ones that will bring the police and fire crews to transport human remains away.

Danforth Elementary and the Texas City High School have been reopened—and Forrest Walker attended his graduation. Like hundreds of others in Texas City, he sometimes sat thinking that his father was going to walk in the door one day, smile at him, talk to him.

A wall has been restored around the vault at Texas City National Bank. Since it reopened, hundreds of people have been coming and waiting in the lobby, applying for loans to rebuild their homes. The national insurance agencies have moved teams of Dallas and Houston adjusters in, and they will be in Texas City for the rest of the year—processing four thousand insurance claims.

At the site of the Monsanto plant, architects in hard hats are watching the rest of the towering, ruined steel girders being yanked to the ground. The roof on St. Mary's Church has been repaired, and people are worshiping indoors. The water supply has been deemed safe to drink. The last wave of DDT has been sprayed on the city.

Still, at least three hundred remain hospitalized. There are long stretches of Texas Avenue, Sixth Street, and the shell roads in El Barrio and The Bottom that are barely rehabilitated. Hundreds of the red CONDEMNED tags, beginning to fray under the brutal hot sun, are still easy to spot on what is left of many homes. Our Lady of the Snows—the anchor in El Barrio—has not been restored. The concrete pilings on the waterfront still look like mountains of dust, steel bars, mesh fence, and left-over steel chunks of the *Grandcamp, High Flyer,* and *Wilson B. Keene*.

They are the immediate visual reminders.

The sixty-three nameless, in limbo several miles away in Camp Wallace, are not seen, but they are on almost every mind.

Curtis has been at Camp Wallace almost daily since the bodies were first taken there. He goes with friends and total strangers who need someone to walk with them through the morgue, making one last attempt to identify their loved ones. Sometimes Curtis has his sons ride in the car with him to Camp Wallace. It is good for them to see how people grieve, how they care for the departed.

Edna and the children never thought Curtis to be blatantly contemplative, but each of them understood, in their own way, that their father was shouldering an entire community's burdens. His youngest son remembers being with his father:

"He was laboring ... he carried that feeling of responsibility for the whole thing and for those who remained unidentified."

Curtis knows the sixty-three have to be buried. Blacks and whites had never been buried in the same cemetery before. Curtis wondered where it was all headed. It was a thought that reared up every day:

What difference could it make now if they're all dead?

Doremus tells Trahan that he will take on the task of finding a resting place for the nameless dead. Doremus begins to wrestle with the hard fact that he has to purchase some land to bury the sixty-three corpses. He is told that he can tap into money from the Texas City Relief Association—the locally run association has been set up to receive the hundreds of thousands of dollars in charity being sent from around the world.

Envelopes are still arriving every day in Texas City, some of them filled with pennies scotch-taped to paper and with the following address: *To the Children of Texas City.*

Even Sam Maceo's series of fund-raisers has gone off just the way he wanted. On Monday morning, April 28, a crowd estimated by the police to be about six thousand people swarmed over the Galveston airport to catch glimpses of Frank Sinatra, Jack Benny, and the others. The city auditorium in Galveston filled up with forty-eight hundred people. Curtis was asked to introduce the governor of Texas, who was going to make one of his infamous blustery speeches. When Curtis stepped to the front of the auditorium, there was lingering, thundering applause for him—more applause than for the governor. An embarrassed Curtis returned to his seat. After the speeches, a lean and hungry-faced Frank Sinatra dropped to a knee, spread his arms, and sang a parody of "Old Man River." Alice Faye offered a version of "Blue Skies." Phil Silvers played the clarinet.

And then the show went on the road, just the way Maceo wanted it, and it sold out in Houston and New Orleans. And, also as promised, after the Texas gigs, Maceo took the Hollywood stars out on his pleasure pier nightclub, The Balinese. On the spot, an oil tycoon from Houston handed Phil Harris $1,000 to hear him sing "That's What I Like About the South."

Maceo is saying that his cavalcade alone will generate $1 million for the Texas City Relief Fund. Other stars outside Maceo's revue are also sending money: Comedian and actor Red Skelton has sent along $3,250; international concert pianists Amparo and Jose Iturbi have sent word that they will stage two piano concerts to raise money. In Manhattan, the stars of the New York Metropolitan Opera are scheduled to stage a benefit performance.

The Texas City Relief Fund will eventually top out at $1,063,000—most of the money will come from corporate donations, including the first donation of $50,000 from the Busch beer family in St. Louis.

There is more than enough to pay for a sliver of the coastal plain, maybe a half acre or a full acre, to bury the unidentified dead.

Doremus, doing the only thing he can think of, begins traveling to the other small communities that ring Texas City. He politely makes inquiries at the city halls, at the chambers of commerce, at the churches.

I am representing an alliance of ministers who are acting on behalf of Mayor Curtis Trahan. Would you have a small plot of land that we might be able to buy so we can bury our sixty-three unidentified dead?

Doremus is told by a friend that the community of Hitchcock, eight miles west of Texas City, has plenty of available, unused land that would be ideal for a small, dignified memorial cemetery. Doremus decides to take a look for himself. The flat piece of the mainland is ideal. He has a good feeling about it; it feels like a place where the unknown dead from Texas City could be suspended in time—where memories would linger. Doremus speaks to people in Hitchcock about the isolated parcel.

They have a question for him: *"Are there any blacks that you are going to bury?"*

Doremus isn't sure what to say. *"Well, yes, there are."*

And, with that, he is told the land is unavailable. No land, in fact, is available to bury blacks. Blacks have never been buried alongside whites, and they never will be.

Doremus is stunned. He decides to tell no one, not even Trahan.

"It was the first really brutal thing that I had happen in my life. It was like a slap in the face. I expected, somehow, at that time, that people would come together in such a way that things like that would never happen. It hit me right in the stomach. There is no guidebook that tells you how to respond, so you just let it go.

"I decided that I would never tell that story to anyone."

FINALLY A TWO-ACRE plot of land is found a mile and a half to the north of the city limits, in a pasture where thin strands of grass bend in the wind and the soil is less spongy. It is a good distance from the waterfront, and if one stands in the center of that piece of earth, the docks and the piers would seem to be a thousand miles away. It is peaceful: only a curving lane passes alongside it, and it feels like a place where solitude will be permanent.

Doremus calls several clergymen and asks them to participate in the dedication of the new Texas City Memorial Cemetery on the third

Sunday of June. A rabbi from Galveston will be there. And Doremus, still shaken from being denied land to bury black people, makes sure to invite Rev. F. M. Johnson—the leading black minister in Texas City.

With the land secured, preparations are made to receive the dead. Up until three days ago, there have still been frantic efforts to identify the remains in Camp Wallace. FBI fingerprint experts pored over any clues one last time.

Now time is up.

Beginning Saturday at 8:00 P.M., a ten-man crew at Camp Wallace had begun preparing the remains. After they are removed from the refrigerated coolers, the remains of each of the sixty-three unidentified are wrapped in a wide burial cloth. A piece of white ducking holds the cloth intact, and, with a crayon, the number assigned to the nameless dead is written on the ducking. The bundles are lowered into identical silk-lined cypress coffins. None of the bodies is intact. None fills the six-foot-three-inch-long caskets. Each casket has a number on it—the same kinds of numbers that had been put on those police traffic tags and affixed to the bodies in the Texas City High School gym. The Camp Wallace crew works through the night.

In the morning, a Houston mortician has arrived with fifty-one hearses. Funeral homes around Texas have donated sprays of flowers. The caskets are loaded onto the hearses and the two-mile-long caravan begins its long trip to the north side of Texas City at 8:45 on Sunday morning, June 22.

The immensity of the moment is incalculable. Some will say nothing remotely like it has ever occurred in the southern half of the United States: a mass burial for blacks and whites who had lived in the most thoroughly segregated ways possible.

Unspoken is the pervasive belief that limbs and organs from several different people have somehow been placed together in some of the caskets. The black and white and Hispanic residents of Texas City are literally being buried together. And now the sixty-three nameless are being sent to their final reward, after weeks of waiting in refrigerated, sterile limbo—and they are being dispatched under the mutual guidance of an Episcopalian, a Methodist, a white Baptist, a black Baptist, a Jew, and a Catholic.

It is muggy; the heat is seemingly swaying along the ground, moving hot and moist among the crowd. Elderly women in veils are being held under the arm, and their children are flipping paper fans in front of their sweating faces. The clouds look leaden, about to burst. A lone airplane is arcing overhead, heading toward Ellington Air Base outside of Houston. The waterfront is far to the west, but there is such a stillness that it seems as if there is a ship's sea horn sounding from Galveston Bay. A preacher moves through the crowd, pressing a copy of the New Testament into every hand—thousands of copies had just arrived from some businessmen in Philadelphia. Curtis has come, and he takes his place alongside the other two thousand mourners. It may be that everyone who has not permanently fled from Texas City—and is healthy enough to venture from a sickbed—has assembled on the patches of toasted grass.

The seven clergymen take their places on a green-felt-covered platform erected in front of a row of bloodred carnations. Alongside the platform there is a spray of pink flowers shaped into a ship's anchor and placed there by some old sailors who had pooled what money they had left. There are small trucks at the edge of the soggy field, waiting to move the damp black dirt across the simple caskets.

Beginning with Nameless Dead Number 243, each casket is carried by a four-man honor guard of longshoremen, firemen, and World War II veterans—black men, white men, and Hispanic men all reaching for the coffin handles. Curtis watches as the sixty-three caskets are placed over six huge trenches, ready to be lowered into the ground after the forty-five-minute ceremony.

This is not the way he and Bill Roach had wanted segregation to end. But maybe Roach knew that this was the only way it could ever happen.

The first to speak is Rabbi Lewis Feegan from Galveston, his prayers crackling out from the speakers placed on top of the sound trucks. Each of the seven clergymen reads a short holy passage. Rev. F. M. Johnson, the Negro preacher, stares out over his first impromptu congregation of white, brown, and black mourners. He had wrestled all night with his choice, trying to find something that would suggest there were painful lessons in the Texas City disaster. He had finally decided to read the ninetieth Psalm. People are bound to hate him, bound to hate the black man's selection—because it says Texas City was being punished by a vengeful God. Texas City was being wickedly punished for all of its sins:

"Thou carriest them away as with a flood. They are as asleep. In the morning they are like grass which groweth up. . . .

"We are consumed by thine anger, and by thy wrath are we troubled . . . thou hast set our inequities before thee, our secret sins in the light of thy countenance. . . .

"Who knoweth the power of thine anger? . . . so teach us to number our days, that we may apply our hearts unto wisdom."

Finally, Frank Doremus steps to the microphone. He reads from the *Book of Common Prayer,* and then he recites the Lord's Prayer. When he finishes, the entire crowd murmurs a response: *"Our Father, who art in heaven."*

The voices of the two thousand mourners sound, someone says, like the insistent, aching roll of the surf along the Gulf of Mexico.

Doremus nods to the casket bearers and, using ropes, they lower the coffins into the shallow trenches. An old man and his son are sinking to their knees alongside one trench, clasping their hands, closing their eyes, and heaving with tears. A middle-aged black woman, her face puffy with grief, is stepping forward and clutching a handful of wilted, white oleander blossoms. She stands over one particular casket, slowly releases the blossoms, and watches as each curling flower petal flutters down.

The tractor drivers and dirt movers watch from the edge of the ceremony. One by one, they flick their cigarettes into the field and start their engines. Each of the trenches will be covered. Sixty-three small granite stones will be placed in neat rows over the graves.

There is nothing on the stones other than the number each of these remains has been assigned. The numbers run, randomly, from 152 to 404.

The Mayor

AUGUST 23, 1947

TWO WEEKS after the explosion, the small-town mayor flew to Washington to appear before the House Appropriations Committee, to beg them to approve a measure allocating a mere $15 million to repair Texas City.

The money will never come.

When he flies home, steps off the plane in Houston, and then drives south to Texas City, it still feels the way it did when he and the other infantrymen were moving through the villages in bombed-out Belgium. There are still those piles of ill-defined, charred pieces of cars, buildings, and furniture clumped on the sides of the road. There are still dozens of homes drooping at sad angles. Businesses that haven't had new glass windows installed have cheap boards covering the frames.

The little things—the things that under normal circumstances possessed no obvious threat—were still reeking of something ominous: the front half of a red tricycle wedged into the ground; a pair of shoes in a ditch; what looks like a long-tailed bird poised in a high tree limb and fighting a breeze . . . but is actually a tattered dress sleeve flapping in the warm wind.

During his brief time away from Texas City, during his trip to visit the powers that be in Washington, he had the time to consider what he had heard at Bill Roach's memorial service. That day the priest giving the eulogy stared down at him—and aimed his comments directly toward Curtis. He had said he had a message from Bill Roach for Curtis Trahan.

Roach wanted Curtis to rebuild an entire city . . . *to rebuild your city of God . . . better, bigger, safer.*

Curtis had no idea if there would ever be any kind of relief from Washington. Many people in Texas City assumed that there would never be any comprehensive aid for a hidden, isolated city where most of the residents were blue-collar workers . . . a city that is run, essentially, by enormously profitable corporations and oilmen whose headquarters are far away from Texas City.

Besides, sending the city a big check would be tantamount to an admission of guilt . . . the King would almost be agreeing that he had done something wrong and that he was sending some money to make up for it. And, besides, it could also set an outrageously expensive precedent—anytime a community suffered an injury of any kind, it could ask the government for special money, special aid, and the government would have to pay it out.

In Washington, D.C., Curtis had moved through the halls of Congress. When he returned home, he had reached a deeply personal crossroads.

That spring, the long-reigning representative from the Ninth Congressional District—which covered both Texas City and affluent Galveston—had taken ill. By the first week of July, he would be dead. Even before he died, the political gears were grinding forward and preparations were being made for all contingencies—including a special election to be held in August.

With only a few weeks until the election, Curtis tells Edna and the children that he is thinking of entering the race for the Ninth Congressional District. He wants to go to Washington again, but this time he would be the one inviting the witnesses from Texas City. He would unleash the flow of relief for Texas City. He would demand to know how an entire American city could be made to suffer the way that Texas City had.

His children, even at a young age, are already aware of their father's firm beliefs:

"*He had, for the greater part of his life, the very strong feeling that government had a responsibility to the people. But at that point, citizens weren't allowed to ask very many questions of the government. There was a certain mystique about the war industry at that time, and no one, including the federal government, was willing to admit that ammonium nitrate blows up.*

"*The federal government at that time didn't really answer to anybody, and the corporations answered only slightly more frequently . . . but for Dad to push for answers . . . that would be very characteristic of Dad.*"

Curtis is a guy who has never gone to college, an oil refinery worker who has drifted into the insurance game, a buck private infantryman blown almost to kingdom come in the Battle of the Bulge, the youngest mayor ever elected in Texas City—and someone who has had the gall to demand that Union Carbide, Amoco, and the others pay some taxes. Then he had become the one person at the center of the most horrific tragedy of its kind in American history. He had talked to the President of the United States, the commanding generals of the U.S. military, J. Edgar Hoover's agents, the governor of Texas. He has even gone for that photograph that the slot-machine kingpins had wanted. He stood alongside Frank Sinatra and all of the most famous celebrities in the world as they descended on Texas, and there was Curtis in the background, taller than everyone else, looking uncomfortable but knowing that he had to be there, it was what the mayor was supposed to do, especially if it could raise some money.

As the political climate quickly shifts, as the local congressional players begin to seriously jockey for position, Trahan receives a call from the Moodys, the wealthiest family along the Gulf Coast of Texas.

Could you come to an important meeting in Galveston?

The Moody clan has made an immense fortune through its ownership of the National Bank of Texas, the American National Insurance Company, and the National Hotel Corporation, which owned properties across the country, including the Hotel Washington in Washington, D.C. Curtis listened as the Moody family urged him to enter the congressional race.

We'll back you. You're visible. You rose to the occasion in Texas City.

Wire stories were moving across the world about Curtis, including the one from United Press International:

"Meet the mayor of Texas City—the community that received the sympathy of the entire world when warlike destruction killed hundreds of people.

"Meet Mayor Curtis Trahan, the thirty-one-year-old infantry veteran, whose grief for his stricken fellow townspeople is hidden only by his grim determination to build a better community over the wreckage and ashes."

Curtis tells the wealthiest family on the Texas Gulf Coast that they are right, he will accept their backing, resign his post as mayor of Texas City, and run for Congress.

He'd get answers and respect for Texas City by going to Washington.

THE RACE IS over before it begins. He never has a chance. Too many big things are outside his control.

After Curtis is cajoled by the Moody clan into running, Clark Thompson also abruptly decides to enter the race. He is married to one of the four daughters in the Moody dynasty. The Moodys own the largest newspaper in the congressional district—the *Galveston News*. They also own the *Texas City Sun*. They have unlimited resources and control the political network in the largest city in the district. Curtis has fallen into a wicked trap, the kind that Lyndon Baines Johnson and others are mastering as they play unforgiving Texas politics:

Galveston will easily throw its votes to Thompson. There are a total of five candidates now. The votes will be divvied, and the race should go to the candidate from the largest city in the district.

Curtis moves ahead and crafts a populist platform. His main campaign aide is the devoted teenager Forrest Walker, who spends hours driving through the city in an old Jeep that Curtis has borrowed and outfitted with a campaign microphone. Curtis goes door to door, trying not to press too hard, trying to suggest that only he can really explain to the people in Washington what the hell had happened that April along the waterfront in Texas City.

Maybe, in the end, he could save some other people around the world. On July 28, a ship that had sailed from Baltimore to Brest, France, exploded in harbor, killing twenty people and wounding five hundred. Ammonium nitrate was still being shipped from U.S. docks, despite what had happened in Texas City—and though some enlightened souls, like the New York fire commissioner, were ordering ammonium nitrate freighters to leave port immediately, there were still many more chances for another major calamity.

Maybe only someone who had lived through the Texas City Disaster could change it all. Maybe it had to be someone like Curtis Trahan.

Forrest Walker admires Curtis. He knows Curtis had roped in the corporations and put a black man on the police force. Curtis has already told Walker he couldn't pay him—but he'd give him a cot to sleep on and buy his meals and let him drive that "official" campaign Jeep. Walker is convinced that Curtis Trahan will be the next congressman from this part of America. Given all that had happened to him, there was no way he could lose.

On August 23, the day of the election, a pounding tropical storm has leaped up, stronger than anyone could imagine. Light poles snapped; water surged out of the gulf and into the streets. Edna complains that the *Texas City Sun,* owned by the Moody dynasty, simply stopped running any stories at all about her husband's campaign. There are other rumors that the Texas City newspaper has muted any coverage of its former mayor in favor of a member of the rich Moody family. In Galveston's paper that day there is a striking full-page advertisement with a roll call of all of Thompson's business supporters.

When the votes are tabulated, Curtis finishes third.

Back at home, though he seems as imperturbable as ever, Edna and the children know it had stung. Curtis simply says this: *"Wrong race, wrong place, wrong time."*

He had no money to fund his race, at least nothing like the wealthiest family for nine counties. Curtis had even returned a $2,000 check that some refinery and union workers had sent him—he said he didn't want any conflicts of interest. He had no organization other than some idealistic teenagers, some people who remembered what he had wanted to do for The Bottom and El Barrio, some friends who believed in him and were willing to go to every single street in Galveston and Texas City handing out his fliers.

In some ways it was easy to view the race on that miserable, storm-soaked day in August as the last bit of proof that the powerful get what they want, and forgettable places like Texas City seem to be filled with aching, unlucky people.

Curtis is weary.

Edna knows her husband better than anyone:

"He was very tired, but he cared about all those people. He'd lost friends and some were never found. Remember all those men he was supposed to go to Houston with that day? They were all killed.

"It was very, very hard for him, but he just felt he had a job to do. That's the way he was. He'd come home just exhausted and try to lie down for a little while. I'd tell him: 'I'll take the phone off the hook for a little while.'

"And he would say: 'Oh, no, you just can't do that.'"

He is no longer mayor of the city in which he had wanted to live forever.

Bill Roach is dead.

He had been right about the one thing he had decided long ago, when he quit his grueling job climbing up the ladders inside the oil refineries:

In this part of the world, there are always going to be storms.

"United States of America, Defendant"

THE SECRETARIES in the U.S. attorney's office in the intimidating federal courthouse in downtown Houston are being handed a thin packet by the tight-lipped U.S. marshal. The usual hum of efficiency will cease as the contents of this special packet are unsealed and examined. They've never seen this before. No one in legal circles anywhere in the nation has. The young clerks from the law library and the interns are leaning in and wanting to know what it's all about. It is a certified copy of a revolutionary lawsuit. It is marked "CIVIL ACTION 787." On the front page, in unmistakably bold print, it reads: *"United States of America, Defendant."*

The perfect irony, as well as the immense ramifications, have not been lost on the well-appointed Galveston lawyer who has signed his name on the space reserved for the plaintiff's attorney.

He knows that today is April 14.

A year ago, the movement of millions of pounds of ammonium nitrate through Texas City had kicked into high gear.

Russel Markwell is like someone who has stepped down from the perfect oil painting of the well-tailored, suave southern barrister. Elegant, clever, and seemingly constantly amused by almost anything anyone has to say, forty-year-old Markwell heads a feisty two-man firm inside the historic, high-ceilinged, marble-tiled Cotton Exchange Building. A broad-faced man, he is a quintessential, almost iconoclastic "islander"—someone, like Elizabeth and Henry Dalehite, who feels right only when he is on Galveston Island and as close as possible to the Gulf of Mexico. For decades he has been the first attorney to come to mind for anyone of any importance locally, from politicians to modern-day pirates, who might need a velvet hammer. For years, Markwell has gently closed his heavy door, pointed the way to a comfortable leather chair, pondered the staggering stories he heard, and betrayed precious few confidences.

Now he has decided to file the first wholesale civil action ever lodged against the United States government—on behalf of the thousands of victims in Texas City.

Since the founding of the nation, Americans had been denied the right to sue their own government for property damage, personal injury, or wrongful death caused by actions of the United States. At the nation's birth, there had been a sense that the government and its leaders had to, needed to, remain immune from the questions and recriminations that ordinary citizens were allowed to level against one another.

Then, a few months before the explosions, Congress finally passed the most ambitious restructuring of American government since the First Congress of the United States. The sweeping bill introduced the modern blueprint for the working structure of the Congress—and it also introduced a related bit of legislation called the Federal Tort Claims Act.

That new law stated that the U.S. government could finally be held accountable—the same way that private citizens could be held accountable for their actions.

As they began to grapple with the profundity of the new law, legal scholars began to say that it was nothing less than the dramatic democratic slaying of "sovereign immunity"—it was the final severing of America's royal roots, the ones that traced back to the ancient, imperial conceit that the men who ruled the land could never be held accountable for their actions.

For decades the conceit was boiled down to a simple, thudding phrase: "The King Can Do No Wrong."

Now Congress had decided that the United States and its leaders were not infallible, not above the law. They could finally be made to answer for the harm, the injury, the death they inflict on their own people . . . and now The King could be judged by his people.

IN HIS BOOK-FILLED and burnished office close to the old cobblestone streets and the sailing ships docked on the Galveston waterfront, Markwell understands exactly what he is embarking upon. His passions are hunting and flying. He enjoys being alone in a small plane, circling over the west and east bays, seeing the hundreds of tiny tributaries, like crooked needles, pointing to the Gulf of Mexico. In his office, surrounded by the framed photographs, the oil paintings, and old gavels, it is manifestly clear.

When the runner walks from his office to deliver the lawsuit to the marshal's office in Houston, it isn't just daring. It is already, in many minds, a brand of mutiny. The word first quickly leaks across Galveston, a place where Markwell knew that you had to work hard to keep secrets from unraveling. In Galveston's dark watering holes, with their nineteenth-century wooden blinds and imported silk curtains, the other Gulf Coast attorneys are sipping their preferred libation—an Old-Fashioned—and trying to reach the essence of what Markwell is doing:

He is trying to sue Washington and the entrenched politicians who presumed to lead the most dominant nation on the planet, the country that has just liberated millions from the death grip of the Nazis and the Japanese.

Privately, Markwell relished the idea of being a trial lawyer from a two-man firm who just might take the whole damned thing all the way to the Supreme Court.

In Houston, one copy of the civil action will go to the local federal attorney. The other two certified copies of the civil action are marked this way:

For Delivery To—Attorney General, Department of Justice,
Washington, D.C.

When the young federal attorney in Houston receives the civil action, he quickly flips through the thirty-three pages. There are seventy-three plaintiffs, including Elizabeth Dalehite, Christine Baumgartner, Rosalie Cedillo, Lupe Garcia, Julia Martinez, Bette Jo Norris, Mercedes Olivares, Judith O'Sullivan, Zenda Meadows, Mattie Westmoreland, Isabel Saragoza, Martha Wood, Christina Westrup. As he scans the list, there is a preponderance of women's names. These are the widows and the mothers caught inside the Texas City Disaster. These are the lead litigants. The attorney hurries to the bottom of the opening page:

"... Each of the deceased persons ... was fatally injured on April 16, 1947, on or near the docks of Texas City ... following a fire in a large quantity of a chemical compound consisting of fine granules of ammonium nitrate. ..."

The attorney flips to page 2:

"... The defendant sold and shipped large quantities of said chemical compound ... to Texas City ... the defendant should have known that said chemical compound ... was inherently dangerous and highly explosive. ..."

Finally, on page 3:

"... Each of the deaths ... was caused by the negligence of the defendant. ..."

ELIZABETH DALEHITE no doubt had to be lingering on his every word.

Markwell has drawn a breath and asked her if she will consider becoming the first test case and the lead litigant, if she will allow him to consolidate hundreds of lawsuits in her name, if she will let him use her name in all the filings in every court setting, if she will trust him when the time came, and it surely would, that she would be called to the witness stand to do battle with the attorneys dispatched from Washington to defeat what they will see as a direct assault on the nation's elected officials.

Markwell wants to know if she will be the first person to pursue seriously a massive civil action against the U.S. government—to go all the way to the Supreme Court, if necessary, to question the infallibility of her country. There will be depositions, testimony, unspoken whispers about her patriotism.

Markwell knew Elizabeth Dalehite and her husband.

Everyone who had gone out on the water, anyone who enjoyed being near the Gulf of Mexico, was bound to encounter the Dalehites. Markwell had studied the endless lists of victims. The Dalehites were substantive, uncomplicated, unfettered by baggage of either a civil or criminal nature—they had a good background; they were hardworking; they were beloved; they had raised their children with due diligence and care.

Elizabeth would be the ideal reminder that the Texas City Disaster was always, in the end, about lives and dreams lost.

Beginning in the days immediately following the explosions, Markwell had been receiving calls from the handful of local Texas City attorneys who had been instantly overwhelmed, inundated, by hundreds of inquiries from the widows, widowers, cousins, grandparents, and friends of people lost in the disaster. There were wills to be untangled, creditors calling, labor union issues to discuss, home and life insurance papers that had been destroyed and now had to be reinvented. There were medical bills that many people had no hope of paying. There were questions about lost wages, back pay, and taxes. There were questions about the legality of seeking medical treatment in other, bigger cities. There were harrowing inquiries about the rights of the families whose loved ones had simply vanished.

For months after the explosion, the numbing bureaucratic maze appeared to have no chance of concluding. And, as in the case with almost every community disaster, the shock and sorrow began to give way to a search for answers—and a welling, righteous indignation.

Elizabeth herself had tried to move on in the last year.

As her son, Henry Jr., had predicted, they buried her husband in the old family plot in the center of the city. It wasn't far from their big, old home on Offat's Bayou. It was something she would always see as she drove to Henry's office and tried to make sense of what he had left behind. Her husband was in that plot, buried alongside the infant, Jacqueline, they had brought into the world. He was there, with all the moss-covered markers and the aboveground mausoleums that some people had built because they worried that another incomprehensible tragedy, a storm perhaps, would uproot the entire earth.

When Henry died, family members and many of her friends encircled Elizabeth. Her generosity over the years now spawned innumerable tender mercies from the people who had heard about her husband's death. There were so many people who seemed heartbroken by her loss. She and Henry had worked, without feeling that it was work and without plotting it, to bond themselves to hundreds of families along the Gulf Coast. Now, for a while after Henry's death, it seemed as if every one of them was at her house, at her church, at her side as she walked along the seawall.

The Dalehite house had been like a reassuring way station for sailors, dreamers, sea captains, deckhands, and fishermen's families. Elizabeth kept half the people in town fed and pretended to be unhappy if they didn't ask for more. Now the collective memory of this part of the Gulf Coast embraced her and held her close.

Weighing what Markwell is asking her to do, she is resolved.

She has had a year to absorb the loss of her husband. She prayed as never before. And she let her mind settle over and over again on the mysteries, the ones that she tried to share with those closest to her, the ones who could possibly understand.

She told her son about her holy statue of Blessed Mary, the way it had been wounded in the same way as Henry, and at the exact moment he walked into the wall of fire.

There are no rational explanations for some things. There is no reason her husband should have died. There is no reason anyone should have died on a beautiful April morning in a place called Texas City. It is the same thing so many people think every day . . . the people in El Barrio . . . in The Bottom . . . in the white neighborhoods with the braids of bougainvillea around the fine houses in the uptown parts of the city. It is the same thing that a black longshoreman like Ceary Johnson will believe as the weeks, months, begin to pass. It is something he will think the rest of his life. . . . *"So many people would have lived if they had only known the truth."*

Elizabeth Dalehite tells the attorney that he can put her name on the front page of Civil Action 787.

The Trial

APRIL 1950

NINETY DAYS after Elizabeth Dale-
hite makes her fateful decision to lend her name to the landmark civil
action, an order of consolidation is filed with the Federal District
Court in Houston.

The document indicates that there are now thousands of parties to
the lawsuit—hundreds of family members, scores of businesses, and
dozens of new attorneys. The consolidated case will bring together 424
suits against the United States brought on behalf of 8,485 people living
and working in and around Texas City.

That number represents the majority of the people left alive. The
lawsuit, in Elizabeth Dalehite's name, is pitting virtually everyone who
lives in one small Gulf Coast town against their government.

Opening arguments are scheduled to commence in the Southern
District of the Federal Court of the United States in October 1949.

Hearing the case will be the legendary seventy-three-year-old U.S.
judge Thomas M. Kennerly.

If Markwell strikes some as a whipsmart, circumspect gambler, then
Kennerly is like the hanging judge, the grim enforcer empowered by
the law and the deep, arcane history of it. He has a hawkish nose, wire-
rim glasses, a shock of sometimes unruly gray hair, a long face, bushy
eyebrows, and the kind of high-collar suits and short ties ideally suited
for someone bridging the nineteenth and twentieth centuries.

He had never formally studied law in any school; instead he had read every law book he could borrow from a lawyer in rural Texas— his mastery of the letter of the law was so immense he was readily admitted to the bar. He catapulted through the ranks, and President Herbert Hoover awarded him a federal judgeship in 1931. Intimidated attorneys who appeared before him would whisper that he was like some immovable object, a stiff-backed Christian who saw the world in unambiguous, black-and-white ways. Staring down from the bench, peering over his glasses, he looked like a grave, stern figure come to life from a nineteenth-century daguerreotype.

During World War II, Kennerly never believed anything as much as he believed the United States of America was on a crusade to preserve basic human dignities. And, as the war raged, he viewed his courtroom as an extension of the United States' effort to maintain freedom. He took any available courtroom moment to do his part:

He especially, readily, agreed to condemn any land that the military needed for the war effort. During the war he handled 254 cases in which the chemical plants and oil refineries were seeking to seize acres of land, build factories, and churn out bombs, airplane fuel, and synthetic rubber. Most of the cases heard by Judge Kennerly were decided in favor of the government, the military, the defense contractors, the oil companies, factories, and plants. He punished individuals and businesses accused of running counter to the war mobilization. He cracked down on anyone questioning the selective service system.

Now, as the landmark Dalehite case approached, attorneys and investigators around the country were craving more information on Kennerly. They were spending days doing feverish spadework, zealously digging through his court opinions for any clues about his inclinations, his politics, his rulings. Quickly, his rock-hard philosophy was obvious. In Kennerly's court, the inflexible edict hammered true and steady: The King Could Do No Wrong.

His new case, CIVIL ACTION 787 on behalf of litigant Elizabeth Dalehite, is more than a test case for the hundreds of other suits rising out of the Texas City Disaster. Lawyers and judges across the nation are already suspecting that Kennerly's decision will figure in the most provocative legal battles imaginable in the future: African Americans demanding satisfaction from the United States for its history of slavery; Native Americans seeking legal relief for the U.S. domination of their independent nations; soldiers suggesting that their generals had negligently led them to certain death.

Kennerly is as inflexible and hard as his hickory gavel, as conservative as anyone who served in the South. He would crush the dangerous notion that the Washington bureaucrats, the politicians, the nation's entrenched leaders had not just neglected ordinary Americans—they were also culpable of willingly maiming and killing them. Kennerly would clearly call it damnable heresy.

As KENNERLY MARCHES into his courtroom in October 1949, he seems even more stern and commanding to the people who have risen in his honor. Without a glimmer of a smile, he gathers his black robes, takes his seat, and fixes his pale eyes on the pack of lawyers and reporters from every corner of the country—including a special squad of defense attorneys sent from the Department of Justice in Washington. Kennerly's face betrays nothing. He seems unmoved by the presence of so many hired guns. He hasn't gone to law school, but he understands the immense gravity of the case. Moving into his seventies, he also understands that he will be orchestrating his own legacy—this will be the most important case he will ever hear.

Earlier, on the day he approved the pretrial plan to make "Civil Action 78: Elizabeth Dalehite vs. United States of America" the single test case for the thousands of plaintiffs, he walked into a similar setting. His courtroom had been as packed then as it is now.

Back then, he stared out over a similar army of attorneys, let a thin smile crack across his face, and thought to himself that every lawyer in Texas was in front of him:

"Be seated, gentlemen . . . it has been a long time since I attended a mass meeting of the Bar."

Today, he is grim and brisk. He had hoped to have this case already heard. The opening of the trial has been repeatedly delayed as he waited for the endless attorneys to finish taking depositions and accumulating evidence.

Since Markwell had unleashed this bombshell lawsuit in April 1948,

Kennerly has received dozens of filings and reports from lawyers—there was a lead team of twelve working with Markwell, there were two lead government attorneys with a pack of assistants.

Kennerly had ordered the rival lawyers to be uneasy traveling companions—to crisscross the country together when taking crucial depositions in Iowa, Nebraska, Chicago, New York, and at various military bases. Months ago, during yet another intense deposition-taking day in a federal courthouse in Manhattan, the raw and frayed nerves were on naked display. The road-weary enemies had come to a slamming halt. They were poised to begin punching. The attorneys for the victims accused the federal attorneys of stalling, doing anything to destroy the case before it went to trial. One of them shouted: *"We mean business. We want to go forward. We're not going to lie around here and take four or five days to take a deposition."*

The U.S. attorney bristled: *"I might say we're in earnest too. That's why we're trying to protect the interests of our client. . . . It is ridiculous to think that we personally are trying to delay this case when we are receiving six dollars a day and are here at a personal sacrifice."*

Now, finally, the lawyers are in court. The hundreds of depositions have been transcribed. A frosty Kennerly begins by demanding that, at the very least, there will be civility in his courtroom. He has always had it in the past. He will have it now.

"I am very anxious to get the case started. I do not want any counsel to have the idea that I am forcing you to trial. All right then, we will proceed."

Kennerly feels the case is easy to digest—even if the attorneys have spent eighteen months accumulating eighty thousand pages of evidence from the White House, the FBI, and the U.S. Army.

It can be broken down into easy mathematics. One simple question: *Is the government responsible for what happened in Texas City?*

Kennerly assumes that the lawyers will argue for ninety days. Then he will retreat to his law books, case studies, and his faithful coterie of law clerks. He will issue a quick and correct decision.

IT IS HALLOWEEN, October 31, and Elizabeth Dalehite is waiting, along with her children, to be called into the courtroom. Before she is called to the stand by Markwell, one of the attorneys at his table asks to speak. Kennerly tells him to proceed. The co-counsel stands and is waving a plastic bag at Kennerly: *"Your Honor, I have the murderer here this morning."*

The lead federal attorney leaps to his feet.

"We object to that and, of course, plead not guilty. If he wants to change the rules of evidence from civil to criminal, we will go along with him."

Kennerly is still not smiling. *"In other words, you have the fertilizer?"*

The co-counsel is still brandishing the bag as if it is a murder weapon. *"Yes. This, Your Honor, is chemically pure ammonium nitrate."*

The government attorney interrupts: *"Very innocuous looking."*

The co-counsel is still giving his speech and admiring the bag in his hands: *"That is the fertilizer-grade ammonium nitrate involved in Texas City. It came out of a sack I appropriated from Pier O within about twenty feet from where the* Grandcamp *crater began."*

He turns to the federal attorney and adds: *"You know where the pier went out and sloped off?"*

The federal attorney is stunned: *"No."*

Markwell's co-counsel says: *"You never went down there? Well, your education is sadly neglected."*

Finally, Kennerly glowers down at the pissing match that has welled up even before Elizabeth Dalehite has been sworn in. He wants it ended. He wants the first witness, maybe the only one who really matters. He wants to see the woman who has been bold enough to lend her name to this precedent-setting test case.

Markwell rises and says: *"We would like to have Mrs. Dalehite in, please."*

MARKWELL WAS RIGHT when he had gone to the other attorneys and said that Elizabeth Dalehite would put the best human face on the Texas City Disaster. In the strategy sessions, clinically dissecting the merits of each of the people they were representing, the lawyers had first talked about someone else. They had talked about other women, other widows. Initially there was a thought that Christine Baumgartner—the widow of the beloved fire chief Henry Baumgartner—would be the ideal "human face" of the Texas City Disaster.

Everyone liked the family; everyone knew them. They seemed to transcend the boundaries in the city—they were liked by people of different colors and backgrounds. Baumgartner, a hero, had died at the epicenter of the explosion. There is a strong chance that he was the first person to die. Baumgartner was literally standing on top of millions of pounds of flaming ammonium nitrate when it ignited and the *Grandcamp* became one enormous bomb. Still, there was a possibility that Baumgartner—who also worked for The Company as a purchasing agent—might be too closely linked to the city officials and powerful people in town.

Other names were examined. The list was obviously long. None of the black names were considered, though Ceary Johnson and Rev. F. M. Johnson both knew that at least 135 black residents had been killed in the explosion.

In the end, there was no one that Markwell wanted more than Elizabeth Dalehite. She is resolute and vulnerable all at once. She seems sturdy but still scarred by what happened. Her arrival at the Texas City waterfront that day was pure fate. There was a phone strike that day; her husband couldn't call down to the docks. He had to be there. She had to take him. They were innocents being pulled toward the hiding fire.

After she is sworn in, Markwell begins to advance her toward her memories of her husband.

"Q. I believe that his father before him had been a seafaring man, had he not?

"A. Yes, he had.

"Q. The Dalehite family had been connected with the sea for many, many years?

"A. Yes, my husband started working on a boat when he was ten years old."

"Q. Why did you go to Texas City?

"A. We had to get our orders. The Seatrain New Yorker *was due in the next day.*

"Q. Mrs. Dalehite, your husband would meet these vessels several miles off Galveston Island out at the bar, isn't that correct?

"A. That is correct."

Then, Markwell comes to the moment that Elizabeth has relived so many times, it is like the dull, gray waves eroding the dark brown sand near her home.

"Q. Now you say your husband got out of the car when you parked . . . just tell us what happened, Mrs. Dalehite."

The entire courtroom appears to be bending to pick up her voice. Kennerly, breathing slow and easy, still shows no emotion.

"A. Well, when he got out of the car I told him, I said, 'Look at the fire over there on the ship.' And he looked and, if I remember, Henry said, 'Come on' and I am sure I said was too tired. Because, you see, I had been driving since about two-thirty that morning, and I just kind of slipped down in the seat . . . and then I always carried along with me a little statue of the Blessed Mother . . . and I used to say prayers while waiting for Henry.

"I always said my morning prayers if I hadn't said them before I left home. I started saying my morning prayers with this little statue in my hand.

"I don't know what happened to Henry after that.

"I didn't see him anymore."

Markwell looks at her.

"Q. Do you have the statue that you had at that time?

"A. I always carry my statue with me.

"Q. What happened to it?

"A. The head was broken off it.

"Q. In the explosion?

"A. Yes it was.

"Q. May I see that, please?"

Before she can hand him the decapitated statue, Kennerly is suddenly leaning forward and interrupting the line of questioning.

"Q. The explosion occurred how long after you parked your car?"

Elizabeth turns to the formidable, gray-maned judge.

"A. Judge, I just couldn't say how long because time meant nothing with me when I was with Henry. . . . I devoted all my time to my husband."

Kennerly, for the first time and one of the few times in a trial that will last seven months, is shaken. He apologizes a bit, trying to string together the right words.

Finally, Markwell interjects and tries to lead her forward to the day when she was at home, still praying that her husband had somehow survived and that he had just gone away for a little while and would be walking in the living room at any minute.

"*Q. Was that the last time you saw your husband, when he was walking from your car to . . . the docks?*

"*A. That is the last time I saw him.*

"*Q. Mrs. Dalehite, you know, of course, that his body was recovered?*

"*A. Yes, I had Henry's body home.*

"*Q. And, of course, you attended the funeral?*

"*A. Yes, I did, very much against the wishes of the doctors, but I did.*

"*Q. You were not permitted to view the body, though?*"

There is absolute quiet in the courtroom.

"*A. Yes, I saw my husband.*"

In the back of the courtroom, her son, Henry Jr., her husband's namesake, is watching. He can't stand to see his mother suffer. Elizabeth goes on:

"*A. He was the most wonderful man in the world . . . he was a good, hardworking man, and he gave us everything in the world we wanted.*"

The reporters and the carefully dressed lawyers are staring at her. In the witness stand it is as if the weary but strong woman is never more alone—and it is as if she is speaking for everyone who lost a husband, wife, or child in what has become known simply as the Texas City Disaster.

THE GOVERNMENT'S CASE centers on shifting blame—including blaming the victims.

For weeks on end, the federal attorneys will produce dozens of witnesses and hundreds of depositions—many of them aimed at incriminating other people for the wholesale carnage. They will also argue hard that there has never been anything inherently dangerous with ammonium nitrate. They begin with the dockworkers.

Julio Luna, the first longshoreman to report smoke in the sweltering belly of the *Grandcamp*, is grilled for hours about cigarettes and smoking—just the way he had been grilled by the Navy when his ship went

down during World War II. Through Luna and the other longshoremen called to give testimony, the government intimates that the dockworkers have always broken safety rules when it comes to smoking
their Lucky Strikes and Camels—and surely one of them carelessly
tossed a still-glowing cigarette onto the bags of ammonium nitrate.

Government attorneys triumphantly produce the final United States
Coast Guard investigation and quote from it:

"The fire in lower No. 4 hold of the Grandcamp *started between 8:10*
A.M., *April 16, 1947, the time longshoremen entered the hold, and 8:20* A.M.
that date, when it was discovered and that it was caused by unauthorized
smoking in the hold."

Now, the government attacks the powers that be in Texas City—
they take aim at The Company, the one business that has worked as a
willing partner with the military and the chemical companies for decades.

Walter Sandberg—who had been pictured on the cover of *Life* with
his head wrapped in a bloody bandage, a fedora on his head, and a
cigar clenched in his teeth—is hammered about the fact that fifty
thousand tons of ammonium nitrate had been transported through his
port in Texas City. The government attorneys seem to suggest that
Sandberg and The Company should have known how to handle
ammonium nitrate—that they were educated, powerful figures who
bore the full responsibility for knowing what dangerous things they
were bringing into their communities.

Sandberg twists and turns, for hours, in the deposition room and
then the courtroom. He says that all of The Company's shipping records, safety memos, and bulletins have been destroyed. He is chal-
lenged about the possibility that the Port of Houston, only forty miles
away, had refused ammonium nitrate shipments. He is repeatedly
asked about the fact that he had been corresponding with officials tied
to the U.S. Army bomb-making plants where the ammonium nitrate
fertilizer was made—why didn't he ask federal officials how to properly store and handle ammonium nitrate?

"Q. . . . did you ever ask anybody's advice or opinion . . . ?

"A. . . . there was never anything furnished us in the way of information
that would indicate that this was hazardous cargo . . . ."

But if the ammonium nitrate was being sent by the U.S. Army bomb plants, wouldn't a reasonable man think it was dangerous?

"*A. This stuff was coming from an Army Ordnance Plant. The papers inside the [rail]car indicated that a captain or officer of the Army had loaded that car, and the fact that it was moving from that plant sealed up by them was all the indication we wanted. . . . There was never any doubt in our minds that we were not handling a harmless commodity.*"

Next, why didn't The Company—the company that built, owned, and operated the docks—have some essential safety items that could saved thousands of victims from injury:

"*Q. Did you maintain a fire tug, a fireboat?*

"*A. No.*

"*Q. Did you maintain a tug or any equipment at all to move out ships that might be on fire?*

"*A. No.*

"*Q. Why didn't you?*

"*A. I beg your pardon?*"

The federal attorney presses harder:

"*Q. Do you regard a fire on a ship of any concern of yours except to protect the property of the Terminal Company?*

"*A. No.*

"*Q. In other words you didn't have an overall safety setup involving a disaster plan, a man who busied himself with nothing else but safety, did you?*

"*A. No.*"

MARKWELL AND HIS co-counsels have scrubbed Elizabeth Dalehite's case down to one powerful idea:

There is a thread linking the White House directly to the concrete pilings on the Texas City waterfront.

The government produced ammonium nitrate in massive quantities for the very bombs that would help win World War II. Then, when the war ended, President Truman had a twin-sided masterstroke. He would keep the domestic economy rolling by converting the wartime ammonium nitrate plants to peacetime ammonium nitrate fertilizer plants. He'd also deliver on his promises to General MacArthur and the foreign leaders—he'd begin shipping ammonium nitrate overseas.

The miracle chemical would resurrect France, Germany, Italy, and Japan—it would keep them fed and blind to the seductions of the Soviet Union.

Dalehite's attorneys will spend days arguing that the nitrate production was overseen by the highest-ranking members of the War Department and the United States Army. The classified White House memos from President Truman, ordering several government agencies to roll forward the billion-dollar programs to feed Germany, Japan, and France, are introduced.

Dalehite's attorneys even force one of the heroes of World War II to testify. Major General Everett S. Hughes, who helped plan the D-Day invasion, and was commander of U.S. troops in Rome after its capture, is commanded to give a statement.

After the war, General Hughes was put in charge of the sprawling U.S. Army Ordnance network—the network that oversaw the production of ammonium nitrate. Now, he is simply asked what he knows about the powers of the chemical compound:

"A. I think that the conclusion to which I have come . . . is that ammonium nitrate has always been regarded as dangerous. . . . There have always been restrictions on the handling of ammonium nitrate, and the shipping of ammonium nitrate."

The World War II general has another question posed to him:

"Q. Did you warn anybody who might handle it, including the railroads and shipping people?

"A. No, I didn't."

Markwell and his co-counsels press on: Ammonium nitrate was shipped, on federal government bills of lading, to ports like Texas City—where even more government agencies, including the U.S. Coast Guard, would be aware of its arrival.

As early as 1941, the Coast Guard had circulated a dangerous cargo list, and ammonium nitrate was listed near the top.

If anyone, anything, knew about ammonium nitrate, it was the federal government.

Markwell and his co-counsels quickly embarked on a droning roll call of nightmares that they said ultimately bred disaster for the 8,485 people in Texas City who were seeking some sort of relief from Washington. They mention the other well-documented ammonium nitrate disasters, especially the deadly one that lifted Nobel laureate Fritz Haber's factory in Germany straight off the ground and almost erased an entire, thriving city.

One of the most important members of the Manhattan Project, the ultrasecret government program to develop nuclear weapons, is also commanded to testify about his personal experiments with ammonium nitrate. Markwell and the others know his résumé is impressive. George Kistiakowsky, born in Russia in 1900, had fought in the tank corps of the White Army until the Bolsheviks seized power. He fled to Germany, moved to Harvard, and was recognized as the leading explosives expert on the Manhattan Project. He had overseen the development of intricate explosive lenses needed to uniformly compress plutonium in order to achieve critical mass. And, when the brilliant atomic scientist explored ammonium nitrate, there were explosions when the heated compound encountered something as simple as axle grease.

Markwell feels it is all so patently obvious.

The government made ammonium nitrate and knew it was the same compound used for bombs. It shipped it to Texas City in bags that contained no warnings—even though every government official who ever produced it, shipped it, and knew about its arrival in a coastal community was aware that it could kill.

Finally, Curtis Trahan is summoned to be sworn in.

After he had been tricked and then overpowered in the congressional election, Trahan opened up a small store in Texas City selling refrigerators and other appliances. The news crews stopped coming. There was a thought that he would write a book about his experiences at the eye of the storm, but nothing ever came of it. Now, he is just about to sell his home. He is thinking of pursuing the American Dream in Las Vegas.

"Q. Before the great disaster, had you ever heard of ammonium nitrate fertilizer or ammonium nitrate?

"A. Not that I recall.

"*Q. Did you know that any of it was moving through Texas City?*

"*A. No sir.*

"*Q. I will ask you this now: At any time before Texas City did any representative of the United States Coast Guard ever call you as Mayor and call your attention to the need of any unusual precautions or any precautions of any sort with respect to the handling of ammonium nitrate through the port?*

"*A. No, sir, the Coast Guard nor any other group, sir.*

"*Q. No one from the Army?*

"*A. No, sir.*

"*Q. Can you tell us about how many people were killed in this explosion?*

"*A. Well, as near as I can recall, it seems it was 398 or 400, right at 400, positively identified dead.*

"*Q. How many unidentified dead?*

"*A. I believe there were 63 unidentified bodies . . . the total was right at 600 that were dead, unidentified, and known missing. Now, when you say 'known missing,' the president of the Terminal Railway is one in that category.*

"*Q. Mr. Mikeska?*

"*A. Mr. Mikeska . . . and I could name a lot more that come in the category of missing dead.*"

"*Q. They were probably transient laborers, people like that?*

"*A. Yes, it could have been.*

"*Q. Had no families there?*

"*A. We got a lot of letters inquiring about people.*"

"*Q. What job did you sort of take on for yourself as Mayor of Texas City?*

"*A. Well, I took on the job of more or less managing whatever situation happened to come up.*

"*Q. By the way, how much sleep did you get the first two days and nights after that?*

"*A. Well, practically none.*

"*Q. Mr. Trahan, when was the time, if you ever did, that you got the first communication from anybody representing the United States Government about ammonium nitrate fertilizer?*

"*A. Well, the only message that I had from the United States Government concerning ammonium nitrate fertilizer was about the second day—the United States Bureau of Mines' man showed me a little Government publication that had a lot of explosives listed in it, and as I recall ammonium nitrate headed the list.*

"Q. When you refer to the second day, what do you mean by the second day?

"A. I mean after the explosion. The explosion was the 16th. It was probably the 18th that this gentleman showed me this publication.

"Q. Was it a thing printed like a magazine would be printed?

"A. Yes, sir. As I recall it was thirty or forty pages, neatly printed on a rather fine finish paper, sort of semigloss.

"Q. Did you at the time think that this Bureau of Mines' man was giving you something that had been printed before the Texas City Disaster?

"A. Oh, yes, sir, I did.

"Q. And it told about ammonium nitrate being an explosive?

"A. . . . I recall that ammonium nitrate was either the first one or right up near the top of the list."

THE CLOSING ARGUMENTS aren't made until the early spring.

Markwell listens as one of his co-counsels sums up:

"These men decided to make ammonium nitrate for use as a high explosive. Those plants were built years before the Texas City Disaster. They were built for the purpose of manufacturing ammonium nitrate to be placed in airplane bombs and dropped on Germany and Japan. . . .

It is the whole government. It is everybody in the organization. Everybody from the President to dishwashers in the cafeteria . . . that is the simplicity of this lawsuit."

There really is nothing more to be added to the argument on behalf of Elizabeth Dalehite, her dead husband, and her family.

The only thing uncertain, thought Markwell, is the domineering, book-schooled country lawyer who has become the most powerful federal judge in his part of the country. In the hallways of his courthouse, they call Judge Thomas M. Kennerly a "black letter" judge—he does everything by the exact letter of the law.

In those same echoing hallways, the betting money is that Kennerly has been as unmoved as the granite that went into the construction of his courthouse. He would simply reject Elizabeth Dalehite's chilling testimony. He would throw the whole damned case out.

No one can blame their own country.

The old judge wasn't prone to blaming his own damned country.

———

KENNERLY IS READY to announce his forty-six-page ruling on April 13, 1950. In Texas City, preparations are under way at that moment for the quiet services to mark the anniversary of the disaster.

Kennerly gravely says that he has carefully combed through twenty thousand pages of trial transcript in the matter of *Elizabeth Dalehite vs. United States of America, defendant.*

He spends a few moments summarizing the history of the case.

And then, he launches into a wholesale, staggering, unforeseeable condemnation of "the defendant"—the United States government:

> *"This Record discloses blunders, mistakes and acts of negligence, both of omission and commission, on the part of Defendant, its agents, servants and employees. . . .*
>
> *"It discloses such disregard of and lack of care for the safety of the public and of persons . . . as to shock one. . . .*
>
> *"When all the facts in this Record are considered, one is not surprised by the Texas City Disaster, i.e., that men and women, boys and girls, in and around Texas City going about their daily tasks in their homes, on the streets, in their places of employment, etc., were suddenly and without warning killed, maimed and wounded. . . . The surprising thing is that there were not more of such disasters."*

Kennerly squashes the government's argument that the explosion is caused by the people of Texas City. The evidence proves to him that powerful ammonium nitrate is exactly what the ancient alchemists marveled at—something that can come to life, something that, when the elements are aligned, can simply spontaneously ignite. He reserves special scorn for the government's argument that ammonium nitrate is not "inherently" dangerous:

"It practically wiped out Texas City ... it was dangerous to manufacture, dangerous to ship, and dangerous to use."

Point by point in the matter of Civil Action 787, the stern, self-educated country judge mercilessly condemns his own country ... the defendant:

"Defendant was negligent in failing to inspect and test ... defendant was negligent in the manner in which it marked and labeled ... defendant was negligent in delivering Fertilizer ... the negligence of the Defendant reached its peak when ... it was shipped entirely across the nation to Texas City, and Defendant did nothing to protect either those handling it or the public against the danger."

Finally, Kennerly puts the blood of its people directly on the hands of the U.S. government. And he does it in his own, inimitable, almost coldly scientific way. In page after page, he mentions brigadier generals, major generals, majors, lieutenant colonels, government scientists, government contractors, the commandant of the United States Coast Guard. He mentions, by name, 168 individuals and says that "in the error and mistake of manufacturing and distributing this dangerous commodity, so many took part that in naming them some will be overlooked or omitted."

For brevity, to pull it all together, he condemns the Defendant:

"It will not do to say that Defendant ... could not reasonably foresee that more than 500 persons would be killed, many persons injured, and that there would be vast property damage....
"Defendant did know."

The old judge has done something profoundly, utterly unexpected. And, in so doing, he has forged himself to the cycle of convergence that is the Texas City Disaster.

Judge Thomas Kennerly's stunning ruling simply says the same thing that Elizabeth Dalehite, a sea captain's widow, believed.

It says the same thing that a World War II veteran and small-town mayor named Curtis Trahan believed. It says the same thing that a black longshoreman like Ceary Johnson forever believed. It is the same thing that Bill Roach, a jangled fool for Christ, an intense priest haunted by his own visions, had believed.

"So many people would have lived if they had only known the truth."

"By Direction of the President
of the United States"

1949–1952

THE WEEK AFTER Kennerly's ver-
dict, Elizabeth Dalehite receives a copy of the same form that hundreds of people are opening up all over Texas City. It is from her
attorney and it has a cautionary sentence underlined:

> *"Dear Client:*
>
> *"As you undoubtedly may have read in the newspapers, Judge
> Kennerly, Judge of the United States District Court for the Southern
> District of the United States, recently held the United States to be
> liable to all persons sustaining damages as a result of the Texas City
> explosion.*
>
> *"Last week Judge Kennerly entered judgment in the test case of
> Elizabeth Dalehite, et al, vs. United States of America. We . . . were
> the attorneys for Mrs. Dalehite and her son Henry G. Dalehite, Jr. who
> were awarded damages in the sum of $75,000 for the death of Captain
> Henry G. Dalehite, Sr. The Dalehite case was selected to be the test
> case by a committee of lawyers representing the plaintiffs, because in
> their judgment it combined many favorable technical elements.*
>
> *"Whether the government will appeal to the Circuit Court and
> to the United States Supreme Court is not known at this present
> time . . . <u>we suggest that you do not make or sign any statements or
> give any information.</u> . . ."*

The total of $75,000 to compensate for the death of Captain Henry Dalehite is broken down this way: For the loss of her husband, Elizabeth Dalehite will be awarded $60,000. For the loss of his father, Henry Jr. will be awarded $15,000. The findings were based on a suggested "loss of future income."

The award, of course, will never compensate for Henry turning away and making that short walk onto the rugged, crowded docks in Texas City. It won't compensate for the fact that their heavily respected business guiding the big foreign ships around the treacherous sandbars, the hidden reefs, and the shifty coastal waters, has simply died along with her husband. Henry Jr., already an outstanding student, wants to follow through on the plans he had shared with his father— no, he wouldn't be following his father to sea; instead he planned to study law and to continue watching over his mother.

Elizabeth will need to find some way to pay for it all, to send her son to school, to keep the old house on the bayou.

In El Barrio, where some people have actually rebuilt their homes, along with Our Lady of the Snows Church, there is quiet but excited talk about the letters being sent by Elizabeth Dalehite's attorney. If the government doesn't appeal, then thousands of others will finally also be getting compensation—and with the money will come the understanding that the workers and families in Texas City had done nothing wrong. The government officials were the guilty party. Now in El Barrio there is hushed talk around the dinner tables that if there is compensation money, then it is, as one father says, "blood money." As with the Dalehites, any judgments will be based on the formula for "lost future income."

All that Elizabeth Dalehite knows is that her small financial award, at least, would perhaps make a dent in her immediate tangle of expenses. An elderly judge in Houston had said her government's leaders were wrong. He had said her husband's life was worth $60,000 to her and $15,000 to her son.

She and her son will never see the money.

Five weeks after Kennerly slowly recites his scathing accusation of the U.S. government, federal attorneys are bustling through the crooked, narrow streets in downtown New Orleans and walking up the steps that lead to the U.S. Fifth Circuit Court of Appeals. Orders have been issued by the attorney general not only to appeal the case with a corps of elite assistant attorneys general sent from the Justice Department—but to also hire hardball Louisiana attorneys who know exactly how the appellate judges think and act.

The unthinkable has happened, and the stakes have been raised to almost infinite proportions. Elizabeth Dalehite's victory has lawyers exploring hundreds of accusations of wrongdoing by the politicians, generals, and contractors.

Hearings are set, dozens of motions are made, the trial is scheduled and rescheduled. Months pass, and it is not until December 7, 1951, at 10:00 A.M. that the six members of the Fifth Circuit Court of Appeals take their seats and await the opening arguments.

The government's new battle plan has already been crafted in carefully guarded meetings in Washington. The case in Texas had turned on the telling testimony, the sad heaviness of Elizabeth Dalehite's appearance on All Saint's Day in 1949. She had an obvious inner resource, something that seemed to rise above the endless, formless stream of innuendo, FBI reports, government documents, railroad bills of lading, and all the other specific bits of evidence. Elizabeth Dalehite, someone who had lived a rich, simple life, had won the case in Texas.

Now the government lawyers will begin their attack by arguing that the new law—the Federal Tort Claims Act allowing citizens to sue the government—doesn't apply to Elizabeth Dalehite and Texas City. It wasn't intended for them. It wasn't intended for what happened to her and anyone else in Texas City.

They will argue that the international interests, the vital foreign policies, of the United States override everything. The global strategies of the United States are of paramount importance—even if it means the death of its citizens. The United States is fighting the growing shadow of communism—fighting the Cold War and the Soviet Union—and that welling war is more important than anything else. This new global battle, this paranoid and grinding and bloody contest that will become the Cold War, outstrips the comparatively insignificant losses in Texas City: the deaths of a sea captain, a fire chief, a boil-ermaker, longshoremen, factory workers, a gaggle of ninth-graders, a public relations man—or even just the breaking heart of a small-town mayor and WWII veteran.

The day of opening arguments, the hired gun, the New Orleans attorney working with the attorneys general, is the first to speak:

"Following the close of World War II, it became necessary to provide food for the starving people of conquered and liberated countries, or in the alter-native, troops to quell rioting. Guns or butter had become a stark reality, and military necessity demanded food for the devastated areas.

"It was found that a ton of fertilizer was equal to seven tons of food. . . . *Idle ordnance plants throughout the United States, it was found, were read-ily convertible to the manufacture of nitrogen fertilizer.* . . .

"By direction of the President of the United States, with approval of the Cabinet under appropriations by Congress, the Secretary of War was directed to use these ordnance plants. . . ."

With that, he has let the government's case settle over the courtroom. The choice is simple.

Either the President, the Cabinet, the Congress, and the generals are right—or Elizabeth Dalehite is right.

ON JUNE 10, 1952, her attorneys have the news.

For years, she has experienced every swing of emotion. For years, like a river painfully and predictably rising, the sorrow has been there. And for the last several years, she has seen the slivers of glass emerging from her skin. It is an awful, common phenomenon. Hundreds of peo-ple who were there that day in Texas City had thousands of needles of glass buried in their skin—and, as the days and months passed, the glass would appear on the surface of an arm, a face, a scalp.

The justices of the federal appeals court unanimously overturn the ancient Texas judge's decision to repay Elizabeth Dalehite. It has been more than five years since Henry died. It has been four years since she and almost eighty-five hundred others had questioned Washington's motivations and actions.

The complete rejection of CIVIL ACTION 787 that day in New Orleans is boiled down to one word:

"Discretion."

The justices determine that the leaders of the United States, beginning with the President, maintain the right to exercise their own "discretion" in matters of vital interest to the nation—even if their plans are faulty, even if their plans are wickedly dangerous to innocent, unsuspecting Americans, "even if some danger were recognized."

The United States needs to make ammonium nitrate. It is of vital, national importance. The Soviet Union is looming. The government is immune from blame. There are clear moments when the King is absolutely right . . . when the King Can Do No Wrong. The United States and its leaders are allowed to take a "calculated risk" with its powerful explosives and with its people:

"The necessity of providing a means of existence to the devastated areas (of the world) might have called for the exercise of discretion as to whether to take a 'calculated risk.' "

The justices conclude their flat, overwhelming reversal by saying that they have heard everything they wanted to hear. Unknowing Americans run the risk of being sacrificed in service to their country's schemes.

No new evidence, no new arguments could have swayed them in the matter of "CIVIL ACTION 787—*Elizabeth Dalehite vs. United States of America, defendant.*"

"All of the evidence had been produced that could be found or produced . . . the judgment of the district court is therefore reversed and judgment here rendered for the defendant."

The King Can Do No Wrong

SUMMER 1953

A YEAR after the hopes of the ordinary heroes in Texas City were crushed, one of the storms of the century roared into the Atlantic. As it moved out over the ocean, it fractured into thousands of thin bands of rain and shrouds of mist that soared into the skies and disappeared forever.

When the armies of the storm first assembled in the heartland, in "tornado alley," they began by leaving 11 dead in Arcadia, Nebraska, and then marched toward Michigan and Ohio. The monster Force 5 winds, clocked at over 250 miles per hour, killed 115 in the blue-collar town of Flint, Michigan, where assembly-line workers had just stepped off their shifts at the massive car factory eventually known as Buick City. It was the last single tornado to cause more than 100 deaths in the United States. The great storm moved steadily east, spawning a tornado in Massachusetts that killed 94 people and became the deadliest New England tornado ever.

At the same time, in Washington, Elizabeth Dalehite's case, as Russel Markwell had once warned her, was going to the United States Supreme Court. Markwell had waited a lifetime for this chance. He had invested more than time and money in the case. He was obsessed—he knew that the final Supreme Court ruling would enter legal texts, classrooms, and lecture halls for as long as law would be taught in the United States.

The simple act of turning down a rutted dockside road on an otherwise glorious morning in April had delivered his innocent client Elizabeth Dalehite to the Supreme Court—still arguing that the government she loved is at fault for the loss of her husband, friends, and all the strangers who had found themselves improbably steered toward an apocalyptic moment in a place called Texas City.

There is no more evidence to present.

The case is beyond new evidence.

Nothing can be added to the eighty thousand pages already accumulated.

Now, it's really a simple question for the Supreme Court Justices:

Are America's presidents and generals personally immune when they use their policies as a shield? Are they automatically free of blame for bloody deeds?

On this day, June 8, 1953, the U.S. Supreme Court has several history-making matters to decide. The justices will rule that restaurants in the District of Columbia cannot not refuse to serve blacks. And then they will turn to the matter of Elizabeth Dalehite and the ordinary citizens of Texas City.

Justice Stanley Reed is selected to deliver the Supreme Court opinion. A former tobacco attorney and solicitor general, the sixty-eight-year-old had been appointed by President Franklin Roosevelt to the court in 1938. Earlier in his career, Reed earned a reputation for fiercely protecting the interests of the presidency—for arguing the constitutionality of presidential decrees. Some whispered that his reward for defending the powers of the White House was an appointment to the Supreme Court.

By now, Reed knows it almost absurd to deny that the government made ammonium nitrate in war plants and was aggressively shipping it around the world. Instead, he neatly lays out the history of why ammonium nitrate was so important to America:

The United States was faced with feeding Germany, Japan, and other nations in order to steer them clear of the Soviet Union. Obeying decisions in the White House, Cabinet officials ordered bomb-making plants to open up again and make thousands more tons of the chemical compound. Congress agreed to pay for it. The U.S. Army's generals were put in charge of the entire operation.

Reed says that no one is arguing that ammonium nitrate isn't dangerous:

"Following the disaster, of course, no one could fail to be impressed with the blunt fact that FGAN (fertilizer-grade ammonium nitrate) would explode. . . ."

The destructive powers of ammonium nitrate are blatantly obvious. But, he argues, so is the chain of command in the United States of America. Government officials who are enforcing important domestic or international policies can't be held negligible. Sometimes, even when death occurs, those government officials must be allowed "discretions."

Reed is saying what everyone in the White House, in the Cabinet and in Congress assumed the ancient Texas judge Thomas Kennerly was going to say years earlier: Government officials really can be infallible. And Reed says there is only one way to interpret the law that allows private citizens to sue their government:

"Congress exercised care to protect the Government from claims, however negligently caused, that affected governmental functions."

In other words, even though the government was negligent, was culpable, awarding legal victory to the people of Texas City would set an enormous precedent—it could freeze government policies, freeze "governmental functions," and probably open elected officials and government employees to endless accusations of harming citizens.

It is possible, as many people in Texas City resigned themselves to thinking, that the lawmakers knew the Texas City Disaster victims deserved some compensation, some help, but were afraid that it would allow thousands of other legitimate demands to be filed against Washington.

Reed's wholesale denial of old Judge Kennerly's cold, mathematical interpretation of the law—and Kennerly's painstaking, detailed finding against all the defense contractors, generals, and politicians—is joined by Supreme Court Justices Harold Burton, Frederick Vinson, and Sherman Minton.

Three dissenting votes are cast by Supreme Court Justices Hugo Black, Felix Frankfurter, and Robert Jackson.

The dissenting opinion is prepared by Jackson, and it is filled with the prophetic echoes that once filled the mind of an increasingly tortured Father William Francis Roach.

Robert Houghwout Jackson, descended from the original settlers of Warren County, Pennsylvania, was one of the brilliant prosecutors of the twentieth century. He was vaulted into the legal pantheon for being able to win a $750,000 judgment against Andrew Mellon, one of the richest men in the world. His fame was secured around the planet when he took a leave from the Supreme Court to serve as the chief U.S. prosecutor during the "trial of the century"—the prosecution of the infamous Nazi hierarchy in Nuremberg. In Germany, Jackson frequently centered his case around the theory that it might be a crime against mankind to plan and wage a war of aggression.

As he weighed the case of a woman named Elizabeth Dalehite, he decided to make an intense, condemning break from the Supreme Court justices who voted against her:

"This was a man-made disaster, it was in no sense an 'Act of God' . . . the disaster was caused by forces set in motion by the Government, completely controlled or controllable by it. Its causative factors were far beyond the knowledge or control of the victims; they were not only incapable of contributing to it, but could not even take shelter or flight from it."

Supreme Court Justice Jackson is reclaiming the connections that the old Texas judge, Thomas Kennerly, had found. The Texas City Disaster is about ordinary lives. Ordinary people. It is about the thousands of people dead or injured or left behind in a forgotten part of America.

They didn't know about the Cold War. They didn't know about ammonium nitrate. They didn't know about the "discretionary" powers of politicians, the defense contractors, and the generals. They accepted that Washington would warn them, protect them. They lived in the burgeoning Age of Chemistry—and faithfully believed that the government and its chemists and its scientists were always acting in their best interests.

Jackson had once argued that the Nazis had committed a crime against society. Now he suggests that some government officials had committed a crime against their own people. And he will also warn that the Age of Chemistry has spawned a "synthetic world," where the building blocks of nature are being dangerously shifted at will.

"This is a day of synthetic living, when to an ever-increasing extent our population is dependent upon mass producers for its food and drink, its cures and complexions, its apparel and gadgets.

"The product must not be tried out on the public, nor must the public be expected to possess the facilities or the technical knowledge to learn for itself of inherent but latent dangers."

During the Nuremberg trials, Jackson argued that Hitler and his Nazis had abandoned their moral obligation. They were the self-deluded, infallible, immune Kings. It was the same thing that Bill Roach had once tried to articulate in his controversial open letter to the *Texas City Sun*. If the politicians did not admit that they were fallible, then they were no worse than Nazis.

The priest, the mayor, the sea captain's widow, the old district judge, and the former war crimes prosecutor had all converged at a common point. Jackson was adamant:

"The Government was liable. If not, the ancient and discredited doctrine that 'The King Can Do No Wrong' has not been uprooted."

In Texas City, there are tears when the devastating news of the Supreme Court's rejection comes down.

The case has lingered for six years. For the eighty-five hundred claimants, it seemed forever.

Life insurance claims have been filed and a few people have received meager death benefits. Monsanto has rebuilt and reopened its plant dangerously close to the old grounds along the waterfront. The scars on the buildings, roads, and streets have, for the most part, been smoothed over. People planting flowers in their yards still touch their spades to buried pieces of Captain de Guillebon's cargo ship.

Children still find unpleasant things along Dollar Point, Rattlesnake Island, and Campbell's Bayou. That fact won't change for decades.

The schools have been rebuilt and reopened. Texas Avenue still serves as the unofficial barrier between the races, though no one seems to enforce it the way they did in the days when the old police chief and city commissioner L. C. DeWalt had shot a Negro five times in the heart. DeWalt's brother Rankin—who had escaped the terror on the waterfront by suddenly, frantically running alongside the black longshoreman Ceary Johnson—will become the new police chief. Some years later he will commit suicide by shooting himself.

The big ships will start docking, and even more oil will be moving through than before. The Company will use its insurance settlements to rebuild and resume control of the docks and the waterfront.

In Texas City, the entire fight against the government has never been an open topic of conversation. Children have been told to not talk about it. If it came up at Lucus's Café or Norris's Café or Clark's Department Store, there would be quiet glares and unspoken commands to abandon the topic.

At the VFW halls along the bleak, debris-strewn coastal roads—at the places where the World War II veterans gathered to swap stories about their days at war—conversations would slam to a halt. The lawsuit, as some people simply called it, has never been something for idle chat or speculation. It's blood money, and it is something big, churning, moving through a tangled system that no one wants to try to understand. And it is, always, something that makes the air leaden with a bittersweet mixture of guilt, anger, and uneasiness.

No one ever really wanted to be in the position of accusing their elected officials of wrongdoing.

Texas City had built the damned country.

No one in Texas City wanted to tear it down.

"Speak Freely"

FALL 1953

MARKWELL TELLS her that he has one last plan, the only plan left, and if it succeeds it will entail her having to testify again. It will mean that Elizabeth will be forced to endure the same questions, and re-create the seconds and minutes of her husband's death all over again—for an entirely new set of strangers.

Rejected at the highest court in the country, he asks if he can have an appointment with U.S. congressman Clark Thompson, representative of the Ninth District. Thompson, of course, is the monied, powerful member of the Moody dynasty—the one who had discovered Henry Dalehite's body in the high school morgue and the one who smashed Curtis Trahan and his teenaged volunteers during Curtis's doomed bid for Congress.

Markwell plans to push Thompson and the powerful minority whip in the Senate, Lyndon Johnson, into sponsoring a bill that will finally send money to the people in Texas City. Johnson has a reelection race welling up. He is the youngest minority whip in the history of the Senate, but most observers assume he wants more. He can use the votes from the blue-collar workers along the Gulf Coast.

Following the backdoor meetings, there are official announcements that there will be a special subcommittee hearing of the House Judiciary Committee in November. It will be held, of all places, in the ornate, lavish Galvez Hotel in Galveston. The hotel is owned by the

Moody family. The hotel is also where the alleged gangster Sam Maceo keeps his penthouse suite—and where Frank Sinatra and the Hollywood celebrities have parties and hold weddings. It is where FDR, Douglas MacArthur, and Dwight Eisenhower like to stay when they visit this part of Texas.

The hearing is under way on November 16 at 10:00 A.M., under the direction of congressmen from Illinois, Maryland, and Massachusetts. U.S. Army brigadier general Claude Mickelwait is also in attendance.

The first witness called is Elizabeth Dalehite's attorney. The congressmen instruct Markwell to take a seat in the witness stand. Markwell asks if, instead, he can stand to speak.

"I would prefer standing, if you don't mind. Being a lawyer, I would feel more at ease."

Markwell has been clipping articles in his office, building a little file of newspaper stories that outline which countries are receiving millions of dollars in foreign aid from the United States. The articles, to him, underscore the fact that the people in Texas City still haven't received any money. He begins:

"We realize that we come to Congress not as matter of right, but more as a moral right that Congress can, if it sees fit, recompense these people. . . ."

As for ammonium nitrate, Markwell adds:

"I think we can prove beyond any shadow of doubt that the Government manufactured it, that it shipped it, and that it controlled it from the day it started the manufacture until the day it exploded on the ship."

Markwell never sits down through the several hours of presentation.

By afternoon, the congressmen adjourn the session and announce that they have decided to reconvene in the morning in the old city auditorium in Texas City—the place where Father Bill Roach had first been taken after the explosion. Elizabeth Dalehite is scheduled to be the star witness.

THE SESSION BEGINS on Tuesday at 1:30 P.M.

Elizabeth Dalehite finds herself waiting in the witness area with the new president of The Company, Walter Sandberg, the man who had taken over control of the docks and waterfront. Also there are the members of the DeWalt family—relatives of the former police chief and city commissioner.

She is sworn in by Representative Edgar Jonas of Illinois. He will lead all the questioning:

"Q. Now . . . *speak freely and tell the committee what you have in mind that you would like to make part of the record. . . .*

"A. *Well, my husband was Captain Henry Dalehite. . . .*

"Q. *Compose yourself and take your time, now. We are all sympathetic toward your cause, but we want to accomplish our purpose here by getting things in the record. We know how disturbing these emotional matters are to you.*"

Elizabeth gathers herself and proceeds. As before, her son is in the back of the hearing room, watching his mother.

"Q. *What date are you talking about now?*

"A. *That was the morning of the 16th.*

"Q. *Morning of the 16th of April?*

"A. *The Texas City Disaster.*

"Q. *What time did you get to Texas City?*

"A. *I am sorry. I can't tell you that because time didn't mean anything to me. . . .*"

Nothing had really changed for her. Her memories were still the same. She moved in sync with her husband and simply made sure he needed to be where he asked her to take him.

"A. *. . . then the next thing I knew, I was blown out of the car, and I—well, I guess I was just like everybody else, just crazy or something, you know. It was just a terrible feeling.*

"Q. *In spite of the fact that you were thrown out of this automobile and hurt yourself, you chauffeured all these other people to the hospital?*

"A. *Yes sir . . . but how I did it I don't know.*

"Q. *. . . you have no property damage here except your automobile?*

"A. *My husband.*

"Q. *Your husband, of course, was killed. . . .*"

"Pleased to Advise"

1954–1955

ON JANUARY 5, 1954, Markwell writes a letter to Lyndon Baines Johnson, who is calculating his plan to move forward as the Senate majority leader:

> *"Dear Senator Johnson:*
> *"I am writing you with reference to some proposed legislation granting relief to the victims of the 'Texas City Disaster' . . . words can hardy express the financial distress that the surviving families of the men killed have suffered for the past seven and a half years. . . ."*

Markwell mails his letter and continues to clip the newspaper. For months and months he adds various items to his file, including a March 15, 1954, article that says a U.S. Army major general had spent $1,200 to build a kennel for his "expensive dogs."

Markwell clips more articles and attaches them to new letters. For months on end, he sends several to Washington, including one to Price Daniel, the other senator from Texas:

> *"Several days ago there was a television newscast showing certain officials of the United States Government paying representa-*

tives of the Japanese government the sum of $2,000,000 as compen-
sation for 22 Japanese who were injured, one of whom later died, as
a result of radioactive burns received because their fishing boat was
in close proximity to an atomic blast set off by the United States
Government. . . .

"It is extremely difficult for the victims of the Texas City Disaster
to understand why the Japanese claims were so promptly satisfied
and theirs have not been."

In the late spring of 1955, he decides to send another letter to Lyndon Johnson, now the powerful Senate majority leader. What has become known as the Texas City Claims Act has been slowly coursing through subcommittees, committees, and debates on the floor of the House and Senate. Markwell's friends in Washington have been telling him that the arguments are contentious—no one wants to pull the trigger and pass a bill that damns the government and rejects the Supreme Court.

"Dear Senator Johnson:

"I am enclosing a clipping from the June 13th issue of the Houston Post *to the effect that we have contributed approximately $100*
billion for foreign aid since 1940.

"It certainly is a pity that the Country can't contribute a little
domestic aid for the widows and orphans of the 570 people that were
killed in Texas City and the 3,500 who were injured."

Several more weeks will pass. Markwell begins to receive telegrams from Johnson's office alerting him to the fact that the bill has taken on momentum. The future is uncertain, but it appears that it may possibly be headed to the Oval Office for consideration.

ON AUGUST 12, at 4 P.M., a Western Union delivery boy bounds up the steps of the Cotton Exchange Building and delivers a special telegram to Markwell's secretaries. It is from Washington and the office of Lyndon Baines Johnson:

Western Union Telegram

1955, August 12

Honorable Russel H. Markwell:

Pleased to advise President has just signed Texas City Claims Bill.

THE SOVIET UNION has detonated its first hydrogen bomb.

And, in its way, the story of the Texas City Disaster needs to be forgotten. It has become a symbol of regression and failure and weakness on the part of somebody or some entity—the President, the military, the corporations, or the American worker. It is an American tragedy best forgotten. It is confined to a small sliver of the American coast, and there are international urgencies that are bigger and more ominous.

On Sixth Street, along Texas Avenue, in the pews of St. Mary's Church, and at the VFW Hall, there is a quiet resignation—forged by the widespread assumption that, after eight years, the President of the United States would still deny Texas City.

Something else will happen at the last minute.

There are rumors that President Dwight Eisenhower loathed the notion of signing a bill that placed blame at the feet of the United States government.

But the President finally signs the Texas Claims Act, and Elizabeth Dalehite, Christine Baumgartner, and the other victims of Texas City begin receiving letters telling them they have 180 days to file a claim.

Even though the old district court judge had once determined that her husband's life was worth at least $60,000—and even though she never received that money—Elizabeth Dalehite and anyone else filing a claim for the loss of their husband, wives, and children are limited to asking for a maximum of $25,000.

Only those who had joined the civil action against the government before April 1950 were allowed to file a claim.

If the victims in Texas City had gotten insurance settlements for the damage to their homes, then that settlement would have to be deducted from anything the government paid out.

The U.S. Army opens a dreary-looking claims center not far from the waterfront. Uniformed military adjusters are sent to live in the area. For two years, they review hundreds upon hundreds of claims. Elizabeth Dalehite and the others fill out the forms in triplicate and carefully include the requested intricate details of how their relatives had died almost a decade earlier.

Russel Markwell talks to the federal officials and arranges it so that Elizabeth Dalehite is the first to receive her check, on a windy spring day when a line of freighters approaches the port and the water from the Gulf of Mexico cracks against the shoreline with the finality of a heavy book being shut.

After reviewing all the cases, the government's adjusters have decided that only 1,394 awards will be made. A total of only $17 million will be granted.

On average, each widow, widower, orphan, or injured victim will receive a government check for $12,195.21.

Good People

APRIL 2001

Elizabeth Dalehite, the Sea Captain's Wife

Henry Dalehite Jr., son of the sea captain and Elizabeth Dalehite, graduated from law school and promptly joined the law firm of Russel Markwell.

On his desk at the Cotton Exchange Building was the task he had been assigned: processing the hundreds of claims for the hundreds of dead in Texas City. For years, it will be his life's grinding work.

As the small checks were finally being dispensed, his mentor Markwell suddenly died. Markwell's old friends knew that he had never been in the case for the money—not that there was much in the end. They said he was an elegant gunslinger, a raconteur emboldened by the chance to finally take a case all the way to the highest court in the land—someone who had finally become convinced that he not only had the fortunes of thousands of ordinary people in his hands, but he also had chance to empower people to take control of their own government. In the middle of the case Markwell knew it would be his legacy.

Elizabeth Dalehite died in Galveston in the spring of 1988, a few days before the anniversary of the Texas City Disaster. She left hundreds of loyal friends and admirers who had known her and her husband. She left behind her two children and eight grandchildren. One of her pallbearers was a grandchild named Henry G. Dalehite III.

Her son knew the toll the entire affair had taken on her. He also knew that his mother should never have been put in the position of having to summon up the kind of courage she had displayed—on the waterfront of Texas City and before the United States Supreme Court:

"It was hard for her. . . . It was hard for everyone. What the government should have done, and it was so simple, was just put warnings on the bags. They just didn't warn people. They didn't warn anyone."

And they took so long to finally send some relief to the small city on the bay.

"A lot of people were in dire straits. They needed help. They were desperate."

Three decades after the explosion Henry Jr. sent a letter to the Department of the Army inquiring whether it was still possible to file a claim on behalf of someone killed in the explosion.

The letter he received back said the Texas City Claims Act had expired in 1962 and "that there is no basis or authorization for presently considering a Texas City disaster claim."

Henry Dalehite Jr. went on to become a prominent state district court judge in Texas, someone recognized for the fairness of his decisions.

Forrest Walker, the High School Senior

Forrest Walker Jr. helped bury his father, Forrest Anderson Walker, the boilermaker at Monsanto, at a graveside service in Missouri.

The teenager graduated from high school a month after the disaster, and he was glad that the senior play had been canceled. The prom was never held in the gym where he had discovered the remains of his father—reduced to a small bundled blanket marked Number 244. The prom was shifted to the city auditorium.

His mother used their life insurance money to send Forrest to college. Then she spent the rest to do what her parents had been too poor to do. She enrolled in the same Texas college at the same time as her son. She became a teacher and married someone whom her husband had known in Texas City.

She moved back to Missouri and died in 1991.

Forrest became a history professor and someone acutely interested in protecting forgotten parts of the peculiar American story that unfolded in Texas City. He is retired, living in Portales, New Mexico. He remembers things:

"I was grief stricken, but not nearly as much as my mother. I don't think she ever really got over it. My parents were devoted; they had quarrels, but everyone does.

"I remember I wanted to hear music on the radio. I wanted to hear jazz, not anything sad. I had to listen to jumpy jazz tunes.

"Of course, I had recurring dreams, too, where we'd find him alive. Over time those dreams went away."

Julio Luna, the Longshoreman

Julio Luna eventually went back to work on the Texas City docks.

Given the fact that he was the first person to alert the world to the rising smoke in the belly of the ship—the smoke that would signal the worst disaster of its kind in American history—his name has been strangely absent from most accounts of that day.

That fact may have to do with the almost aggressive way the city fathers of Texas City—the ones who followed Curtis Trahan—refused to draw attention to the disaster.

That fact may also have to do with Luna's fears—fears he would be accused of starting the fire that almost erased an entire city, fears that he would be singled out for extra invective because he is Hispanic.

Luna now lives not far from where his family once lived in a community of Mexican American immigrants taking shelter in the old, abandoned Santa Fe boxcars. He is one of the dwindling numbers of dockworkers, longshoremen, and merchant sailors who gather at the low-slung local VFW hall to sip beer quietly and talk about their days and nights fighting for their country. He has answered many questions.

"The FBI talked to me over and over and over again. Just like when I was in the navy and our ship went down. All kinds of questions."

Henry Baumgartner, the Fire Chief

Harold Baumgartner, the twelve-year-old who had watched his father be the first person to die in the Texas City Disaster, returned to school, and the family was consoled by what seemed to be the entire city. In several memorials, for several years, his father was saluted as a true hero and a beloved fixture in the community.

At the small, solemn memorial service held annually at the Memorial Cemetery for the unidentified dead, a Texas City fireman in a crisp, polished uniform rings a sturdy fire bell. The echo chimes out through the hot air, disappearing in the general direction of the miles of refinery looming along the waterfront. As he rings the fire bell, he recites the names of each of the twenty-seven fallen members of the department.

He begins, as always, with Fire Chief Henry Baumgartner.

Mike Mikeska and Walter Sandberg, The Company Men

Mike Mikeska, the often beloved and respected president of The Company, is presumably buried among the unidentified dead in the Memorial Cemetery—presumably laid to rest alongside the French sailors, the black longshoremen, the children from El Barrio, and the fallen firefighters. No one is sure. His many acquaintances from the community, his business, and in politics would remember him in various written and spoken tributes over the years.

His daughter Beth, who was among the last in the family to see him and who continued to live in Texas City, endured several interviews with FBI investigators seeking clues to his whereabouts—anything that could help identify him.

Like everyone in Texas City who lost a loved one, she thought about her father almost daily—and the fact that he had vanished without a trace.

Walter "Swede" Sandberg became president of the railroad and docks after his mentor Mike Mikeska disappeared. When Sandberg appeared on the cover of *Life* magazine in the week after the disaster—his head bloodied but a big cigar still defiantly clutched in his teeth—he became a sort of symbol for Texas City.

Under his direction, and after several years, Texas City resumed its status as one of the busiest petrochemical ports in the United States. During the years of court proceedings, Sandberg sat on the witness stand longer than anyone else in Texas City. Until he died a few years ago, there were lingering accusations that he should have known about the dangers on the waterfront—and that, even long after the disaster, the Texas City waterfront continued to be a time bomb waiting to explode.

Kathryne Stewart, the Mother and Widow

Kathryne Stewart left Texas City with her two children and returned to Galena Park—the small town where the family had lived before they arrived in Texas City. With the loss of her husband's income from the oil refinery, she resumed her teaching career. She helped her sister-in-law Ivy write one of the truly extraordinary memoirs of the Texas City Disaster; the work was published by a so-called vanity press in 1962.

John Hill, the Chemical Engineer and Baseball Hotshot

John Hill, the hotshot baseball player who skipped a chance to play for the New York Yankees—and then found himself appointed, on the spot, the deputy mayor who would stare down the general of the U.S. Fourth Army—spent several more years working for Union Carbide.

He kept exhaustive files and notes of his days in Texas City—including bags of letters addressed this simple way: "John Hill, Deputy Mayor, Texas City, Texas."

More than almost anyone in Texas City, he spent years poring over the intricate details of how the community had responded to the tragedy. He forced attention on the need for every municipality in America to have a point-by-point "emergency management plan" to cover every possible disaster.

The Texas City Disaster—and how people responded to it—has become required reading for any student of emergency management in the United States. Almost every federal, state, and city law for responding to enormous disasters—man-made disasters, tornadoes, hurricanes, fires, and floods—has been based on what happened in Texas City.

Two years after the disaster, Texas City banned the movement of ammonium nitrate in any form or container on the waterfront. Texas City now has the emergency equipment and manpower of a city five times its size.

Florencio Jasso, the Merchant Seaman

His wooden shack, among the closest to the waterfront, was destroyed. His mother's leg was amputated. Almost all his neighbors had their homes knocked down or a loved one killed.

Florencio Jasso moved west of town, but he watched friends and neighbors in El Barrio do the only thing they could: they began rebuilding their bungalows back in the shadow of the Monsanto chemical plant. Together, the members of the fractured Mexican American community restored the wooden Catholic church. There were months of vigils, wakes, and burials. And when the compensation checks from the U.S. government arrived there was very little celebration.

A typical reaction in El Barrio came from someone who had lived near Jasso. During the disaster, Jesus Jimenez had pulled his daughter Ernestine out of the wreckage, but she died on the way to the hospital—and when a check came for her death, he assembled his family. He reached into his pocket. "I just wanted to tell you that I've got a check here for $25,000 we received for Ernestine's death. It is very precious, but it is blood money and very sacred. We have to treat it as such."

Jasso stopped shipping out on merchant marine ships and applied for one of the hundreds of jobs at the new Johnson Space Center up the highway near Houston. He watched as the chemical plants began buying one home after another in El Barrio and knocking them down.

By the 1990s, El Barrio was completely gone. Today there is a uniform collection of narrow streets with nothing on them but an occasional shrub or half-dead tree standing in strange, neat rows—rows that once defined someone's yard or home.

There is one tree there that appears vigorous. It is a magnolia planted by Teodoro Garcia—the patriarch of the community who lived to be 110 years old. The other famous Garcia in Texas City was Trinidad Garcia, the man who once rode with Pancho Villa and then wound up working for forty years as a janitor and lawn cutter for the men who ran The Company. He died in 1983. His children took jobs in the chemical plants and the oil refineries once the color lines came down. Several members of his extended family eventually developed virulent health problems—cancers, kidney failure, and other deadly illnesses. One of his sons, who worked for years in the Amoco plant, died of cancer.

The waterfront was completely rebuilt, and more refineries and chemical plants went on-line than ever before. The port and the Texas City Terminal Railway remained privately held—owned by Union Pacific and Burlington Northern. It became the eighth-largest port in the United States. The population grew to fifty thousand.

Several observers allege the region has had a disproportionate number of leukemias, cancers, bone diseases, and glandular disorders. Environmental activists in Texas have dubbed it "Toxic City" and have cited Monsanto, Amoco, and Union Carbide for allowing Texas to have the greatest percentage of toxic emissions in the country. Local officials say the numbers are wrong and that Texas City is as safe to live in as any place in America. They point with pride to the fact that it has been named one of the country's "All American Cities" by the National Civic League in 1997.

The Texas City tin smelter, once the largest in the world, is closed and heavily guarded. The United States Environmental Protection Agency has declared it one of the most lethal toxic sites in America, and it has been placed in the "Superfund" category. Meanwhile, Florencio Jasso and some others wonder if secret work was done in Texas City in the 1950s to produce enriched uranium for the American nuclear arsenal. Other residents say Agent Orange was also produced in Texas City. Agent Orange, used as a herbicide during the Vietnam War, has been linked to a number of fatal diseases. Among the companies to produce Agent Orange was Monsanto. The company and others settled a class-action lawsuit in 1987 for $180 million.

Recent studies by environmentalists allege that people living near Texas City face risks higher than the goals set by the Clean Air Act. The allegations are incorrect, say many in Texas City.

A half century after ammonium nitrate almost ended the existence of Texas City, some environmental studies allege that the number one pollutant in the community is ammonia. Leading the list of the other alleged major pollutants in Texas City are nitrate compounds.

Texas is the leading state in the United States for the production of ammonium nitrate fertilizer.

Ceary Johnson, Longshoreman and Dockworker

They are almost all gone now.

Ceary Johnson is the last of the original black longshoremen still alive in Texas City.

He buried all of his friends who died that terrible day. He listened to his neighbor, the Rev. F. M. Johnson, deliver dozens of eulogies—to the first massive mixed-race service to be held in the South. The preacher would stay in Texas City, rebuild his home and his church. With the color line permanently broken, he would be invited to speak at the annual memorial services held in the cemetery for the unidentified dead.

Over the years, Ceary Johnson watched the refineries expand as never before. Union Carbide was growing. And Monsanto was back and bigger than ever. Within a year, the plant had reopened. Monsanto had been awarded at the time the largest single-loss claim and settlement in the history of American insurance.

Johnson opened up his own pool hall a few blocks away from the plant. He kept his eye open for the dwindling number of jobs that called for strong hands and a strong back on the docks. But with the docks still healing, the flow of cotton, timber, and sugar slowed down in Texas City. A way of life was ending.

A few dozen of his neighbors have hung on in The Bottom. The Booker T. Washington School building is still there, but it is closed to students. The handful of homes still there are perpetually engulfed by the constant, heavy noise and smoke from the chemical plants and refineries. People there, and in many parts of Texas City, still grab their children and hit the floor when they hear a car backfiring, the boom of a fighter jet, or some early Fourth of July fireworks.

For years there have been other explosions and massive chemical leaks along the Texas City waterfront:

In 1986, a fireball soared off the docks and was seen by people eight miles out on the water. The blast, in a boxcar loaded with volatile buta-diene, killed one man. A lawsuit in the case granted the dead man's ten-year-old son $12 million. The boy was to be paid $2,400 a month for the rest of his life, with an annual 3 percent raise. The boy was also to be given four payments of $50,000 each to pay for college.

It was a far cry from the average, onetime payments of $12,000 awarded for the Texas City Disaster.

The next year, on the exact April anniversary of the disaster, the Texas Center for Rural Studies issued a study saying that nothing had changed in Texas City. The environmental organization announced that "we'd all like to believe that a tragedy of that dimension would have taught us a very strong lesson. If anything, our lives and the future of the state may be in more danger than ever before."

A few months later, four thousand residents of Texas City were forced to evacuate after a massive release of thousands of pounds of toxic hydrofluoric acid from the petrochemical complex. The poison-ous cloud moved over the city, and 1,037 people, including many chil-dren, were hospitalized for varying illnesses. Whole stands of trees, bushes, and shrubs were quickly defoliated.

Federal investigators said that Texas City had narrowly missed hav-ing thousands of people die, thus averting the American equivalent of what happened in Bhopal, India. A massive chemical leak at the Union Carbide plant in Bhopal is generally considered the deadliest chemical disaster in history.

In 1992, another plant worker died in an explosion inside an isomer-
ization unit. On Mother's Day in 1994, there was another massive
chemical release that drove more people to evacuate or be hospitalized.

In April 1996—three days after the annual memorial service for the
Texas City Disaster—the entire city was put under alert, and several
people in The Bottom were evacuated. A self-proclaimed international terrorist had gone to a pay phone on Interstate 45, electronically
garbled his voice, called 911, and said that a series of bombs were
going to detonate at 6:00 P.M. in the middle of the chemical plants. He
said that he had planted seven devices but that only four were real
bombs.

U.S. Coast Guard officers ordered six freighters to flee the port in
Texas City and then closed down the ship channel to traffic. Parents
raced to remove their children from schools. The FBI sent agents, just
as they had in 1947.

Bomb experts eventually discovered two devices, including one on
the grounds of the rebuilt Monsanto plant. Both were deemed to be
"dummy bombs." It was the anniversary of the bombing of the federal
building in Oklahoma, where, of course, Timothy McVeigh had used
ammonium nitrate for his death-dealing mechanism.

April continued to be the cruelest month.

In April 1998 there was yet another toxic release rising up from the
petrochemical complex and blowing over the waterfront and the
homes near the docks. More people were hospitalized, complaining of
rashes, breathing problems and headaches. Skeptics said that people
were spraying ammonia in their eyes and claiming to be injured.

All that Ceary Johnson knows is that he is a survivor.

He finally moved uptown, away from the waterfront, and up to
where the flowers seemed to prosper on the north side of Texas City.

One day he opens the screen door to his home in the part of town
that blacks once were never allowed to visit. He lets a stranger, someone curious about the Texas City Disaster, come inside. The old man
gingerly lowers himself into a well-worn, cloth-covered easy chair. He
is ninety-two, slightly stooped but still muscled from a lifetime of
work on the waterfront. His eyes are the color of runny eggs. When he

grips the stranger's hand, he doesn't let go for a long time, and his long fingers are like slender but sturdy sticks from one of the ancient live oak trees swaying with the uncomfortably hot winds moving through his neighborhood.

"Some people think they know what happened," Johnson says, his scratchy voice rising. "They don't know. You . . . don't know. Sometimes people don't want to know the truth. Do you know that? Some people don't want to know the truth, so they'll go with a lie."

Johnson's room would be dark except for a thousand brilliant pinheads of light spearing through the door screen.

There are two faded black-and-white photographs on his small end table, each of them showing a handsome, vigorous Ceary Johnson in an impeccable suit and surrounded by other young family men who appear hopeful for the future. Near his television there is also a small pile of yellowed newspaper clippings, documents, and files. They are chipped and brittle around the edges. They are the old stories about the Texas City Disaster.

On what started as an immaculate day in April 1947, the predictable world vanished for many people like Ceary Johnson—and it was easy to find Christians who could only fathom it as a final act of retribution, a direct punishment from on high for the enormous complexities of a place in America called Texas City.

"You know what I thought?" Johnson asks. "I thought it was Judgment Day."

He pauses for a second and stares to see if the stranger can even comprehend it.

"For some reason, God didn't take me."

Just like the stranger who abruptly arrived to ask him about the disaster after all those years, there are lines of moisture moving down his forehead and sunken cheeks. His long-sleeved shirt is blemished by spreading blossoms of sweat. Outside, the tree leaves are chattering all the way to the polluted waters linking Campbell's Bayou to the old oyster reefs of Shoal Point and then to the northern bay area called Dollar Point . . . the place that earned its name because slaves could be bought there for a dollar a head. The temperature gauge on the door of the clapboard grocery store down the block reads ninety-six. The humidity is mired in the high nineties. There is a sheen on the visitor's skin, and his pores are widening from the moisture in the air.

"Sometimes people can bring something awful on themselves," says Ceary Johnson, his eyes beginning to moisten with tears.

The old man is perfectly still, and the only sound in the house is his breathing. He has closed his eyes, squeezing them shut.

"So many people would have lived if they had only known the truth."

Bill Roach, the Priest

For years, the rumors and sightings never stopped.

People in Texas City swore that they had seen Father Bill Roach jauntily walking through the dregs of the poor neighborhoods along the waterfront.

It was, in fact, his twin brother, Johnny—still looking for some semblance of understanding about what happened the day his brother died. In the late 1980s, Johnny died; friends said it was because he had never really recovered from losing his brother.

The people who had accepted the notion that Bill had indeed died years earlier on the waterfront, told differing versions of his death for fifty years: He died in the second explosion, the one when the *High Flyer* erupted. He had given the last rites to dozens of people. On his deathbed, he asked about the well-being of everyone else but himself.

The radical hospital he and his brother had envisioned for Negroes eventually opened and became a bedrock institution for minorities in the state of Texas. Black doctors had volunteered during the Texas City Disaster—and made national headlines ("Race Doctors Treat White Patients for First Time"). Those black doctors were all eventually hired to work at Roach's hospital.

Not long before he died, Bill had also followed the rumored instructions of the nun whom he and his brother had once visited—the one some say had told him that there would be blood flowing in the streets. She had asked him to open up a retreat, somewhere safe and far from Texas City. Bill cajoled church elders into opening up a cloister deep in rural Texas. On one visit to the cloistered nuns, he shared his vision—naming specific friends of theirs who were going to die in Texas City. The nun who bears his name—and who had been the daughter of the man who brought the Amoco refinery to Texas City—entered that sheltered community fifty years ago. She is still there, and sometimes she tells visitors of her deep regard for Father Bill Roach, that sometimes she can feel his presence in the room.

There are still several people pursuing the possibility that Father Bill Roach can be canonized as a saint by the Catholic Church.

Curtis Trahan, the Mayor

One of Edna's aunts had a connection at the casinos in Las Vegas, and that's how he went to the desert.

Curtis went to take his chances with the gangsters in Nevada.

He was a good manager; he knew his way around crises. He got a job as the maintenance man at one of the downtown casinos for Benny Binion—the notorious gun-toting Texas gambler who owned the Horseshoe, invented the World Series of Poker, and helped rule the city in the desert.

At first, after he was tricked and crushed in his race for Congress, Curtis had tried to make ends meet by opening a little appliance store in Texas City. Curtis's little business—selling washers, refrigerators, and even record albums—quickly failed. The place was out of the way, up on the second floor of the Odd Fellows Hall and down the block from where he once had an office in City Hall.

Sometimes, at night, he wondered why everything had happened on his watch. He wondered why they didn't just pull the cocks on the ships and let the freighters sink into the oily waters off the coast of Texas City.

Somebody must have thought there was something on the ships worth saving. Something worth saving more than Texas City and its people.

The mayor who followed Trahan was an anti-union man. Trahan's shared vision with Father Bill Roach—to completely refurbish El Barrio and The Bottom—was never realized. The communities remained segregated, unimproved, and still within a rock's throw of the chemical plants and oil refineries.

Fifteen years after the explosion, some small public housing units were built in The Bottom. In a perverse acknowledgment of history, the complex was named the Grandcamp Apartments.

The families who lived there have reported, for years, a variety of strange breathing problems, skin growths, and coughs.

Curtis stayed in Las Vegas for five years, moved into the food side of the casino operation, and then relocated to various cities across the country. He and Edna helped run a hotel right on the Gulf of Mexico in Biloxi, Mississippi. He ran hotels in Chicago and then in Columbus, Ohio. The last one he ran was in the deep south of Texas, just north of the border with Mexico.

In the middle of it all, he and Edna drifted apart and divorced. They made sure their two children went to college, and both their boys enjoyed outstanding academic careers.

Curtis eventually remarried and then entered a retirement center a few miles from Mexico, where the muddy, sluggish Rio Grande begins to empty out into the Gulf of Mexico. Health problems cloud some of his memories of what happened in Texas City. On some days he is painstakingly lucid. On other days, he is less clear—or less willing to talk about what happened.

Physically unbent by the years, he graciously receives visitors; he loves to dance and he cooks when he can. He speaks glowingly about his children and grandchildren, though he sometimes forgets their names and the places in which they live.

When a stranger comes one day and asks him about his proudest moments in Texas City, he does what people remember him doing years ago. He runs one of his large hands over the spot where his leg had almost been taken off in World War II. He looks down at the floor and says he didn't really do too much worth remembering.

"People credit me? They do? Well . . ."

He says he remembers naming the first black man to the police force. He remembers wanting to improve living standards down on the waterfront. And his face breaks into a warm, knowing smile when the name of his old friend Bill Roach is mentioned to him. Those memories have not escaped him.

He says he has no photographs or mementos from his days as mayor of Texas City—he slowly got rid of everything that captured him at the epicenter of one of the greatest tragedies in American history.

He had lived through killer hurricanes and been in the Battle of the Bulge when the buzz bombs came down like black meteors. He had sailed on the *Queen Mary* and cried with the other soldiers when they heard their commander in chief had died. He shook hands with Frank Sinatra, talked to the highest-ranking generals in the nation, and corresponded with the President of the United States.

Curtis leans forward and smiles at the stranger who has come to visit.

Curtis is now eighty-five years old. His whole life, it seems, has always been surrounded by storms. He is quiet for a very long time. Everything and everyone seemed to converge at once in Texas City.

Now, hopefully, there shouldn't be any more storms.

"I always loved Texas City, and I thought I'd be there forever.

"Good people . . . hardworking people . . . lived there."

Acknowledgments

Down by the docks, down by the stained waterfront, the people in Texas City don't ever ask for much—certainly not for a book to be written about them. These are workers, hard workers, people who have stopped being afraid of the unforgiving grind that awaits them when they walk into the metal forest of refineries, chemical plants, and factories.

The heroism, dignity, and quiet resolve of the Texas City survivors are testimony to the fact that the backbone of America is strong. These are the hero workers, the people who for decades have done the thankless toil that produces things, that makes things, that has a real-world cause and effect. They are patriots, in the best sense of the word.

In Texas City it is not too much of a stretch to say that there is precious little time for idle melancholy. In Texas City, you must be alert to do your dangerous work deep inside the chemical plants, the factories, the refineries. There are gigantic charts, near the dock, that document the last time there was an accident, a fatality. Your mind cannot be clouded by doubt or, say the critics, distracting dreams.

That said, the mistakes in this work are mine and mine alone, and I apologize for the discomfort they may cause for any of the survivors of Texas City. The work is culled from years of research and dates to the time when I first heard about the forgotten disaster in 1980—on a day when I had innocently followed some advice to drive to Texas City and try some fishing off the dike.

Curtis Trahan spent hours with me, in person and on the phone. His thoughtful, patient, and kind family shared hours of insights with my research team. The Trahans, all of them, including his former wife, Edna, and his wonderful children, are the quiet heroes, the best kind. They are justly proud of the way they behaved in the face of tragedy, and they have never sought acclaim or attention. The help they afforded was vital, important, and kind.

Again, thanks must be extended to the people who have extensively chronicled the short, brilliant life of Father Bill Roach with letters, extensive biographies, correspondence, photographs, and archival records. These are wonderful, warm-hearted people committed to the memory of a saintly man who fought for social justice, for human decency, and who truly cared for everyone in Texas City—regardless of race, color, or creed: Lisa May, archivist with the Catholic Diocese in Galveston, Texas; Rev. James Vanderholt; Thelma Avant; Elva Rogers; Doris Pire; Bernice Smith; Sister William; Rev. Frank Doremus; Monsignor Jamail; and Rev. Thomas Culhane.

Ceary Johnson was a blessed soul and a wise guide who allowed me—a total stranger—to come into his living room. He took me by the hand and led me through the fog of history along the Gulf of Mexico. Forrest Walker's sensitive words, memories, and work were exceedingly important. Florencio Jasso is a brilliant man who understands history more than most historians; he spent his valuable time showing me and my researchers where to begin looking. Julio Luna was willing to endure more questions. The exceedingly smart and tough John Hill took time out to spend hours with me, going over the specific details of what happened, sharing his critical personal files, and getting to the truth. His spontaneous work to save Texas City is a tribute to how everyday heroes live among us. Rev. F. M. Johnson gripped my hand and told me I had a lot to learn—and he was right. The closest friends and admirers of Father Bill Roach brought him to life.

Susie Moncla, the dazzlingly intelligent head librarian at the Moore Memorial Library in Texas City, has spent years making sure that her archives were the final resting place for the American tragedy in Texas

City. This book would not exist if not for her unflagging efforts. Her assistants, especially Joanne Turner, were always quick to help. A key, learned historian in Texas City also provided invaluable assistance— Merriworth Mabry had the wisdom and foresight to chronicle the life and times of her city in a way that no one else had done. Through the library, I met Jesse Ponce—who worked in the nearby petrochemical complex for decades and was also a wonderful, good-natured historian doing diligent work to keep the memories alive.

The entire Texas history staff of the excellent Rosenberg Library in Galveston, including the gracious and thoughtful Casey Green, the wise Anna Peebler, and energetic Shelly Kelly, must be thanked for allowing me access to their wonderful collection. They pointed the way to many papers, letters, and files that had rarely been seen by researchers. And, when new material magically arrived from the dead storage files of ancient law firms, they were quick to process the yellowed documents and make them available to me. Their courtesies and professionalism will always be remembered.

The authors of five exceedingly important volumes provided endless leads, theories, thoughts, and information used in this book. Foremost is Dr. Hugh Stephens, the astute former political scientist from the University of Houston. His precise, painstaking work on the Texas City Disaster is, in a word, excellent. The depth of his research is both profound and courageous—Dr. Stephens was the first student of the Texas City disaster to question all the usual assumptions. I had the good fortune of meeting him and talking about Texas City. He readily shared his extensive knowledge, and, as I delved into my research, I constantly saw his detective's tracks ahead of me. Credit also goes to Elizabeth Wheaton, Ron Stone, and Ivy Stewart Deckerd for compiling immensely useful works—sadly long out of print—in the wake of the disaster. Much of the account of what happened to public relations officer Basil Stewart is derived from the wonderfully accurate narrative work done by his sister Ivy. Veteran Houston newsman Stone's ninety-three-page examination of the Texas City Disaster was enormously valuable; it shows his usual solid work, and he conducted some of the last interviews with key figures in the tragedy.

The gifted French writer Jean-Yves Brouard, one of the finest maritime historians and writers in the world, is thanked for offering me his advice and for his penning the masterful *Le Drame du Grandcamp*—the splendidly detailed French-language perspective of the doomed sailors who came to Texas City. His essential work provided the basis for my understanding of the men who were steered to Texas City by a rising tide. His book deserves to be published in English. My thanks to a gifted man with a classic brand of existential humor—Walter Moore—for translating portions of Brouard's book.

At the National Archives, Barbara Rust allowed me access to many of the eighty thousand pages of original evidence, testimony, and depositions related to the lawsuit filed by the citizens of Texas City against the United States of America. Sifting through the dozens of boxes was a transport back in time—and into the office of President Harry Truman, the mind of General Douglas MacArthur, the plots of Lyndon Baines Johnson, and the lives of the blue-collar workers on a waterfront in Texas. I remember holding the original, crumbling Texas City Police Department log sheets in my hand—seeing the fevered, panicked scrawls that were written at 9:12 A.M. when the world ended in Texas City.

The word "EMERGENCY" was scribbled in bleeding ink.

My faithful researchers are to be thanked.

First is W. Michael Smith, who reenlisted to work with me again after he so ably helped on my biography of President George W. Bush. Mike endured more than any researcher should—my mood swings, demands, inconsistencies, and rants. He bent but, I think, did not break. His work (with me and with a galaxy of varied stars including Gail Sheehy, Dan Rather, CBS, PBS, *Vanity Fair*, and the BBC) is proof to me that there is no finer researcher in America.

Next is veteran investigative reporter Jordan Smith, who also helped with research for my biography of President Bush. She showed her usual energy, commitment, and care in seeking out sources, interviews, opinions, and those buried documents in our National Archives.

Kristen Kelch, Martha Nelson, Tina Brown (and her husband, Harry Evans), Bob Wallace, and Ellen Kampinsky are to be thanked.

Newsweek National Affairs editor Tom Watson, writer Mike Geffner, author Bill Crawford, photographer Robert Seale, Louie Canelakes (who owns the finest tavern-restaurant in Texas—"Louie's" in Dallas) and veteran editor Laura "Buster" Jacobus are to be thanked for defining what true, honest friendship means. For advice and kind words now and in the past, I must thank David Maraniss and David Broder at the *Washington Post*; R. W. Apple, Frank Bruni, and Frank Rich at the *New York Times*; Buzz Bissinger, Mario Puzo, Dan Rather, Vanessa Valencia, brilliant author-journalist Patrick Beach, Bill Lodge (one of America's most courageous investigative journalists), Jo Virgil, Maury Maverick, Jr., author Becky Chavarria-Chairez, the soulful Steve Levin, the generous Joel Draut, Barbara Belejack (the fiercely intelligent coeditor of *The Texas Observer*, the vital, crusading magazine that Father Bill Roach would no doubt have loved), Maryln Schwartz, and Bob Compton. They are shining counterpoints to the lonely, petty, self-loathing newspaper editors and newspaper bureau chiefs of the world.

My agent David Hale Smith is thanked for his hard work and advice (as, of course, is the fine staff at his literary agency). My editor Mauro DiPreta has been patient, encouraging, and brilliant. I'm lucky to have met him and to have worked with him—he saved my ass innumerable times and did it with a sense of humor. He is a good man. His editorial assistant Joelle Yudin is talented, humane, and a joy to work with.

My mother, Tess Minutaglio, a woman who was born in 1917 and raised in a foundling's home, is like the women in Texas City—the ones who survived with a bottomless well of inner strength. I thought about my mother as I wrote about people who spent their lives doing hard work. My other family members—Linda and David Smeltzer; Martha Williams and her husband, Tom; Emily Williams; my brothers Robert, Thomas, Frank, and John—have been loving and kind in many ways . . . and in many ways that I am sure I will never know.

My children, Nicholas Xavier and Rose Angelina, are, were, like sweet balm. When the heaviness of so much misery in Texas City pressed down on me, they reminded me of my blessings. As I wrote about the stoic families in Texas City, the ones who lost their loved ones, I learned to step out of my writing cave and pull my children close. My wife is, as one of my friends once said, a saint merely for being married to me. Holly was the most beautiful, talented human being I had ever known when we met many years ago.

She still is.